Topics in Positive Biology:
The Centenarian Lesson

Editors:

Calogero Caruso
Department of Pathobiology and Forensic and Medical Biotechnologies, University of Palermo, Palermo, Italy

Sonya Vasto
Department of Molecular and Biomolecular Sciences and Technologies, University of Palermo, Palermo, Italy

IMMUNITY & AGEING

Editor-in-Chief: Calogero Caruso

Published by

IMMUNITY & AGEING

Immunity & Ageing is published by :

BioMed Central Ltd
Floor 6, 236 Gray's Inn Road
London WC1X 8HB, UK
T: +44 (0) 20 3192 2000
F: +44 (0) 20 3192 2010
E: info@biomedcentral.com
E: immunity_ageing@unipa.it (editorial enquiries)

Immunity & Ageing can be found on the web
at the following address:
http://www.immunityageing.com

Open Access

All articles in *Immunity & Ageing* are published under BioMed Central's Open Access Charter (http://www.biomedcentral.com/info/about/charter): they are freely available online and archived in full in PubMed Central (http://www.pubmedcentral.nih.gov/), and all users may download, store, or redistribute them in their original form as long as the original citation and bibliographic details remain with each article. BioMed Central is committed to maintaining open access for all research articles that it publishes, both retrospectively and prospectively, in all eventualities.

For further information about the journal, please see the description available on the website (http://www.immunityageing.com)

Disclaimer
Whilst every effort is made by the publishers and editors to see that no inaccurate or misleading data, opinions or statements appear in this publication, they wish to make it clear that the data and opinions appearing in the articles and advertisements herein are the responsibility of the contributor or advertiser concerned. Accordingly, the publishers, the editor and the editorial board and their respective employees, officers and agents accept no liability whatsoever for the consequences of any such inaccurate or misleading data, opinion or statement.

Indexing services
Immunity & Ageing is indexed by CAS, Citebase, Embase, Google Scholar, Index Copernicus, OAIster, PubMed, PubMed Central, SCImago, Scopus, SOCOLAR, Zetoc.

Front cover images
The pictures depict a group of Sicilian Centenarians from Monte Sicani, one of the places in the Western part of Sicily characterized by a high presence of oldest old. "Monti Sicani" encompasses the area between the cities of Palermo and Agrigento from North to South and between the city of Caltanissetta and Trapani from West to East. The territory is characterized by a hilly area of clayey sandstone or pasture and a mountain area above 900 m which consists of pelagic limestone rocks of the Mesozoic era. This area is characterized by olive tree agriculture, which tolerates a large range of soil conditions, preferring a neutral to alkaline soil type. Courtesy of Sonya Vasto.

Back cover images
Courtesy of the Editors

Topics in Positive Biology:
The Centenarian Lesson

Contents

Introduction

Long-lived individuals live beyond the life expectancy (the average number of years a person can be expected to live) of their cohort; hence, longevity is a dynamic phenomenon, as the range of age whereby a subject can be defined as 'long-lived' changes depending on its birth cohort – the survival curves, in fact, mutate with time. In other words, life expectancy changes with time by modifying the range of the age at which subjects can be defined as long-lived. This is not surprising as, unlike the physical phenomena, biological phenomena have to be contextualized. However, to be a centenarian is a longevity index, and in the Western world the number of centenarians has exponentially grown in recent decades. This increase is the result of an epidemiological transition that has occurred over the last 150 years, during which time the average life expectancy has undergone a dramatic increase. This is a result of improved diet and decreased exposure to infections with a consequent reduction in inflammatory processes, as well as the progress in medicine that has reduced the main causes of mortality in the elderly. All this enables an increasing number of individuals to reach the maximum life span that, for humans is around, 110-120 years, with the Guinness World Record held by Mrs. Jeanne Calment who died in 1997 at the age of 122 years 5 months and 14 days. Due to the gradual decline in mortality (1-2% per year) in octogenarians, which began around in the 1960s in all industrialized countries, in the past half century the number of centenarians in these countries has increased by more than twenty times.

The geographical distribution of centenarians is not homogeneous, however, and over the years many geographical areas with a high prevalence of centenarians have been described. In the 1970s three ethnic groups with a high concentration of centenarians were identified in Ecuador, Georgia and Northern Kashmir. However, the absence of a reliable registry office prevented the confirmation of the existence of extraordinary longevity in these areas. Conversely, in the Japanese islands of Okinawa a high prevalence of centenarians has been well documented with a ratio of 42.7 centenarians per 10,000 inhabitants (for comparison, in Italy the ratio is 2.4). By 2050, most of the centenarians are expected to live in China, USA, Japan and India. Currently, France, Japan, Spain, Italy and Canada are the countries with the highest concentration of centenarians, while the number is decreasing in Scandinavian countries. This is correlated with the degree of support from their offspring, which is best in France, Japan, Spain, Italy and Canada and minimal in Scandinavia.

Understanding why some people can live much longer than others is a complex matter. This complexity arises as new, exciting and also contradictory scientific knowledge emerges. The phenomenon of ageing and longevity is, in fact, extremely complex and requires an integrated approach, going beyond purely biological studies to incorporate studies of various social sciences, including the demographic, historical and anthropological sciences.

Centenarians thus become a model for studying the biological determinants of ageing and longevity, as distinguished from those in which the "usual" ageing has led to a lower life expectancy. They are a good example of how genetic, epigenetic, biological, cultural, socio-economic and stochastic determinants have been combined in an optimal way, allowing them to live a long and healthy life - in other words, they represent an *in vivo* model of successful ageing. It is therefore a selected population of individuals who have escaped or survived the diseases that most frequently occur in old age, such as cardiovascular diseases, cancers, diabetes and Alzheimer's disease.

What is amazing in the papers that describe the clinical condition of centenarians is the presence of 25-30% of persons employed in activities of daily living, and 88% that were reported to be functionally independent up to 90 years. A considerable proportion of centenarians have completely evaded dementia and, for those affected, its clinical manifestation appears only in the last years of life.

The hypothesis of compression of morbidity seems to be valid for centenarians regarding functional disability. The onset of disease-related disability can be postponed to the last years of life, but they can also live for 100 years and over with a history of age-related diseases due to a functional reserve that allows some to survive to old age despite the presence of diseases that would otherwise be associated with a significant risk of premature mortality and disability.

Rather than making diseases the central focus of research, "positive biology" seeks to understand the causes of positive phenotypes, trying to explain the biological mechanisms of health and well-being. The purpose of this book is to realize, through a "positive biology" approach, how to prevent and/or reduce elderly frailty and disability. To this aim, the relevance of genetics and lifestyle in the attainment of longevity will be discussed.

Calogero Caruso and Sonya Vasto
University of Palermo, Italy

Caruso et al. Immunity & Ageing 2012, **9**:5
http://www.immunityageing.com/content/9/1/5

IMMUNITY & AGEING

EDITORIAL

Open Access

"Positive biology": the centenarian lesson

Calogero Caruso[1]*, Giuseppe Passarino[2], Annibale Puca[3] and Giovanni Scapagnini[4]

Abstract

The extraordinary increase of the elderly in developed countries underscore the importance of studies on ageing and longevity and the need for the prompt spread of knowledge about ageing in order to satisfactorily decrease the medical, economic and social problems associated to advancing years, because of the increased number of individuals not autonomous and affected by invalidating pathologies.

Centenarians are equipped to reach the extreme limits of human life span and, most importantly, to show relatively good health, being able to perform their routine daily life and to escape fatal age-related diseases. Thus, they are the best example of extreme longevity, representing selected people in which the appearance of major age-related diseases, such as cancer, and cardiovascular diseases among others, has been consistently delayed or escaped. To discuss the relevance of genetics and life style in the attainment of longevity, five papers mostly focused on Italian centenarians have been assembled in this series. The aim is to realize, through a" positive biology" approach (rather than making diseases the central focus of research, "positive biology" seeks to understand the causes of positive phenotypes, trying to explain the biological mechanisms of health and well-being) how to prevent and/or reduce elderly frailty and disability.

Keywords: Ageing, Frailty, Longevity, "Positive Biology"

Introduction

During the last century, life expectancy at birth rose by a remarkable 30 years in Western countries and in Japan, initially because of reductions in infant, child, and maternal mortality and then because of declining mortality in middle and old age. So, during the past century, humans have gained more years of average life expectancy than in the last 10,000 years: we are now living in a rapidly ageing world. The sharp rise in life expectancy, coupled to a steady decline in birth rates in all developed countries, has led to an unprecedented demographic revolution characterized by an explosive growth in the number and proportion of older people. In 1900, about 40% of babies born in Western countries were expected to live beyond age 65. Today in these same countries more than 88% of all newborns will live past age 65 and at least 44% will live beyond age 85. The number of people aged 60 years or older exceeded 635 million in 2002, and is expected to grow to nearly 2 billion by 2050. The proportion of people aged 60 and over stands about 1 in 4 in many Western countries as well as in Japan. Should the present trend continues, this ratio is expected to reach 1 in 3 by 2050. So, many countries have rising ageing populations and are facing an increased prevalence of age-related diseases and increasing healthcare costs, since the rapid rise in older people is accompanied by an increase in the number of people with chronic age-related diseases. However, the improvement in public health has reduced the principal causes of mortality in the elderly, allowing an increasing number of individuals to reach the maximum lifespan age. Indeed, around the 1950s, in all industrialized countries, the progressive decline of mortality in oldest old people has increased, so that the number of centenarians has augmented about 20 times [1-6]. Nowadays it is reasonable to assume that the total number of centenarians is more than three hundred thousand people worldwide [7].

The increased ability to reach 100 years in industrialized countries over the last 150 years most likely reflects a rise in life expectancy as a consequence of improvements in diet and a reduced exposure to infection and inflammation [1]. In favour of diet as a modulator of longevity, the Elderly Prospective Cohort Study identified a reduced overall mortality among the elderly consuming a modified Mediterranean diet in which saturated fatty acids were substituted for monounsaturated ones [8] and

* Correspondence: calogero.caruso@unipa.it
[1]Department of Pathobiology and Medical and Forensic Biotechnologies, University of Palermo, Corso Tukory 211, 90134, Palermo, Italy
Full list of author information is available at the end of the article

the zones of Sardinia and Sicily characterized by male longevity are characterized by close adherence to Mediterranean diet [9]. The traditional Mediterranean diet provides about 40% of calories from fat, mostly monounsaturated and polyunsaturated fat [8]. Concerning Japanese centenarians, traditional Okinawan diets provide about 90% of calories from vegetables carbohydrate, therefore it is low in calories but nutritionally dense, particularly with regard to vitamins, minerals, and phytonutrients [10]. Concerning inflammation, the reduction in lifetime exposure to infectious diseases and other sources of inflammation, the cohort mechanism, has been suggested to contribute to the historical decline in old-age mortality, suggesting long-life pathogen burden as the most important factor for age-related inflammation. Accordingly, some studies have linked an individual exposure to past infection to levels of chronic inflammation and to increased risk of heart attack, stroke, and cancer [11-14].

The extraordinary increase of the elderly in developed countries underscore the importance of studies on ageing and longevity and the need for the prompt spread of knowledge about ageing in order to satisfactorily decrease the medical, economic and social problems associated to advancing years for the increase of the subjects which are not autonomous and are affected by invalidating pathologies [15].

Recently, it has been pointed out that most biomedical research should be termed 'negative biology', because the study of disease is its central heart, reflecting the prevalence of pathology-oriented negative biology, so focusing on the causes of pathology. By contrast, the Author invites to focus on a different approach, named "positive biology". Rather than making diseases the central focus of research, positive biology seeks to understand the causes of positive phenotypes, trying to explain the biological mechanisms of health and well-being [16]. In particular, concerning our topic, this means to understand why some individuals, i.e. the centenarians, have escaped neonatal mortality, pre-antibiotic era diseases, and fatal outcomes of age-related diseases, so leaving more than 100 years [17,18]. Investigating the biological mechanisms underlying centenarian longevity, therefore, shows an interesting conundrum for positive biology. The knowledge grew out of this approach could allow to modulate the rate of ageing providing valuable information on the lifestyle to achieve healthy ageing. In addition, studying centenarians might supply important indications how to build up drugs that can slow or delay ageing, with benefits for those who are more vulnerable to disease and disability [19,20].

The model of centenarians is not simply an additional model with respect to well-studied organisms, since the study of humans has revealed characteristics of ageing and longevity as geographical and sex differences, role of antigenic load and inflammation, which did not emerge from studies in laboratory model systems and organisms. So, scientists have focused their attention on centenarians as optimal model to address the biological mechanisms of successful ageing [17,18].

Centenarians are equipped to reach the extreme limits of human life span and, most importantly, to show relatively good health, being able to perform their routine daily life and to escape fatal age-related diseases. Thus, they are the best example of extreme longevity, representing selected peope in which the appearance of major age-related diseases, such as cancer, and cardiovascular diseases among others, has been consistently delayed or escaped [17,18].

The ageing process and longevity are multi-factorial events. Genetic, epigenetic, stochastic and environmental factors seem to have a crucial role in ageing and longevity. As well known, life expectancy is a familial trait and longevity is determined by different factors. Epidemiological evidence for a genetic component to variation in human lifespan comes from twin studies and family studies. By comparing life span in twins, researchers have found that approximately 25% of the overall variation in human lifespan can be attributed to genetic factors, which become more relevant for extreme longevity [21-24]. Conditioning factors, that arise in the first part of life (socio-economic state of parents, education and month of birth, which has been found to reflect the environmental conditions during the prenatal and early postnatal period), account for another 25% of such variability; life circumstances at adult and old age (including socio-economic status and medical assistance) may account for about the remaining 50% [25]. In this context, the study of centenarian offspring, a group of healthy elderly people with a familiar history of longevity, might help gerontologists to better identify the correlation between genetic profile and hope of a healthy ageing. Previous studies have reported that centenarian offspring, like their centenarian parents, have genetic and immune system advantages, which reflect a minor risk to develop major age-related diseases, such as cardiovascular diseases, hypertension or diabetes mellitus as well as cancer [26,27].

The series

To discuss the relevance of genetics and life style in the attainment of longevity, five papers mostly focused on Italian centenarians have been assembled in this series with aim to understand, through a "positive biology" approach, how to prevent and/or reduce elderly frailty and disability.

As it is known, healthy ageing and longevity in humans result from a number of factors, including genetic background, favorable environmental and social factors and chance. So, in their paper Montesanto et al., discuss the role of epidemiological, genetic and epigenetic factors in the variation of quality of ageing and lifespan, including the most promising candidate genes investigated so far. Epigenetic modifications indicate the sum of heritable changes, such as DNA methylation, histone modification and miRNA expression, that affect gene expression without changing the DNA sequence. So, they outlined some recent advance in the epigenetic studies of ageing, as epigenetics, a bridge between genetics and environment, might explain many aspects of ageing and longevity.

In their review, Ferrario et al,. point out that the genetic origin of exceptional longevity and the more recently observed environment-driven increase in the average age of the population could possibly be explained by the same genetic variants and environmentally modulated mechanisms. The potential overlap of hits for environmentally and genetically mediated predisposition for extreme longevity in centenarians is highlighted by the association of genetic variants of genes that regulate, or that are regulated by, nutrient metabolism. They conclude that the adoption of innovative study designs combined with novel genetic platforms and innovative statistical methods hopefully will bring to the identification of new intervention points at which to modulate ageing and the diseases of ageing.

In their review, Balistreri et al., report their data gathered for over 10 years in Sicilian centenarians. Based on their findings, they suggest longevity as the result of an optimal performance of immune system and an over-expression of anti-inflammatory sequence variants of immune/inflammatory genes. The data from Sicilian investigation add another piece to complex puzzle of genetic and environmental factors involved in the control of life span expectancy in humans, showing a complex network of trans-inactive genes able to influence the type and strength of immune responses to environmental stressors, and as final result, conditioning individual life expectancy.

The paper of Vasto et al., pays attention on the modifiable lifestyle factors such as diet and nutrition to achieve extension of health span. Previous data reported that in Sicily, the biggest Mediterranean island, there are some mountain regions where there is a high frequency of male centenarians with respect to the Italian average. The present data show that in Sicani Mountain zone there are more centenarians with respect to the Italian average. In fact, in five villages of Sicani Mountains, there where 19 people with age ranging from 100–107 years old, on the total population of 18,328 inhabitants. So, the centenarian number was more than 4.32-fold higher the national average (10.37 vs. 2.4/10,000); the female/male ratio was 1.1:1 in the study area, while the national ratio is 4.54:1. Unequivocally, their nutritional assessment showed a high adherence to the Mediterranean nutritional profile with low glycemic index food consumed.

In their paper, Davinelli et al., focus on dietary patterns of centenarians and nutrient-sensing pathways that have a pivotal role in the regulation of life span. They point out that the realization of healthy longevity is possible but to achieve a longer and a healthier life, increased attention must be placed on lifestyle choices, particularly the diet. To date the main dietary intervention that may retard the ageing process is calorie restriction and a typical example is the Okinawan population in Japan. Many of the genes that act as key regulators of lifespan also have known functions in nutrient sensing, thus called nutrient-sensing longevity genes and variant associated to longevity have been described.

Conclusion

Success in increasing longevity in laboratory organisms has shown that ageing is not an immutable process [28]. Hence, the time has come to get more serious about the effort to slow human ageing or to age successfully. On the other hand, if ageing is combined with extended years of healthy life, it could also produce unprecedented social, economic, and health dividends [29,30]. Thus, particular attention has been centered on centenarian genetic background, immune system and life style.

Centenarians, despite being exposed to the same environmental conditions as members of the average population, manage to live much longer; moreover, as a consequence of demographic selection, centenarians have a compression of morbidity and mortality towards the end of their life-span [31]. On the other hand, it seems that long lived individuals harbor genetic risk factors for age-related diseases as recently underlined also by genome-wide association studies data, reporting as very long lived individuals share the same number of risk alleles for coronary artery disease, cancer, and type 2 diabetes than younger controls from the same population, thus suggesting that human longevity is not compromised by the cumulative effect of a set of risk alleles for common disease [32-34]. These studies support the existence of buffering mechanisms operating in the determination of human longevity, probably through the presence of favorable genotypes contrasting the deleterious effect of age-related disease genes: as a result, the frequency of deleterious genotypes may increase among individuals with extreme lifespan because their protective genotype allows disease-related genes to accumulate [35]. A better understanding of the functional genes that affect healthy longevity in humans may lead to a rational basis for intervention strategies that can delay or prevent age-related diseases. However, with the exception of APOE and FOXO3A variants, none of the many candidate genetic variants tested to date have been consistently replicated across populations. This is possibly on account of differing environmental stimuli generating inconsistent demographic pressures, making results, as a consequence, irreproducible [36].

Regarding immunological aspects, studying centenarian offspring reveal that several B cell immune parameters are better preserved than in age-matched controls, and together with their genetic background could contribute to their healthier ageing. This suggests the idea of the "familiar youth" of the immune system [37,38].

Concerning life style, it is out of doubt that healthy centenarians live surrounded by a solid support network of friends and family. However, diet plays a key role in successful ageing. Specific dietary factors that may be involved include a high intake of fruit and vegetables, and in the property of phytochemicals to activate specific nutrient sensing pathways. Centenarians have a very high intake of phytochemicals in the diet. All plants contain these natural compounds and the elderly have significantly lower levels of lipid peroxidation and they suffer less free-radical-induced damage [39-41]. From a scientific perspective, a particular diet able to delay ageing may help to identify new molecules to extend and ameliorate lifespan, opening new opportunities for drug discovery and companies working in nutrition and pharmacology. As recently stated [16], drugs able to mime the effects of calorie restriction might postpone most age-related diseases. In this manner, we could get a much greater health bonus for elderly than overcoming any one definite age-related disease. It has been calculated, in fact, that the slowdown of ageing rate by seven years might reduce the age-specific risk of death and frailty by about half at every age [16,30].

In conclusion, the development of strategies that will lead to the extension of healthy life and that would result in slowing the rate of ageing may be part of the new paradigm for the medical sciences that is the 'positive biology'.

Competing interests
The authors declare that they have no competing interests.

Authors' contributions
CC wrote the paper. All authors edited the paper and approved its final version.

Author details
[1]Department of Pathobiology and Medical and Forensic Biotechnologies, University of Palermo, Corso Tukory 211, 90134, Palermo, Italy. [2]Department of Cell Biology, University of Calabria, Rende, Italy. [3]IRCCS Multimedica, Milan, Italy and University of Salerno, Baronissi, Italy. [4]Department of Health Sciences, University of Molise, Campobasso, Italy.

Received: 5 April 2012 Accepted: 23 April 2012
Published: 23 April 2012

References
1. Oeppen J, Vaupel JW: **Demography Broken limits to life expectancy.** *Science* 2002, **296:**1029–1031.
2. Olshansky SJ, Ault AB: **The fourth stage of the epidemiologic transition: the age of delayed degenerative diseases.** *Milbank Q* 1986, **64:**355–391.
3. Butler RN, Miller RA, Perry D, Carnes BA, Williams TF, Cassel C, Brody J, Bernard MA, Partridge L, Kirkwood T, Martin GM, Olshansky SJ: **New model of health promotion and disease prevention for the 21st century.** *BMJ* 2008, **337:**149–150.
4. Vasto S, Caruso C: **Immunity & Ageing: a new journal looking at ageing from an immunological point of view.** *Immun Ageing* 2004, **1:**1.
5. Salvioli S, Olivieri F, Marchegiani F, Cardelli M, Santoro A, Bellavista E, Mishto M, Invidia L, Capri M, Valensin S, Sevini F, Cevenini E, Celani L, Lescai F, Gonos E, Caruso C, Paolisso G, De Benedictis G, Monti D, Franceschi C: **Genes, ageing and longevity in humans: problems, advantages and perspectives.** *Free Radic Res* 2006, **40:**1303–1323.
6. Troen BR: **The biology of aging.** *Mt Sinai J Med* 2003, **70:**3–22.
7. United Nations "World Population Prospects: The 2008 Revision"
8. Trichopoulou A, Orfanos P, Norat T, Bueno-de-Mesquita B, Ocké MC, Peeters PH, van der Schouw YT, Boeing H, Hoffmann K, Boffetta P, Nagel G, Masala G, Krogh V, Panico S, Tumino R, Vineis P, Bamia C, Naska A, Benetou V, Ferrari P, Slimani N, Pera G, Martinez-Garcia C, Navarro C, Rodriguez-Barranco M, Dorronsoro M, Spencer EA, Key TJ, Bingham S, Khaw KT, Kesse E, Clavel-Chapelon F, Boutron-Ruault MC, Berglund G, Wirfalt E, Hallmans G, Johansson I, Tjonneland A, Olsen A, Overvad K, Hundborg HH, Riboli E, Trichopoulos D: **Modified Mediterranean diet and survival: EPIC-elderly prospective cohort study.** *BMJ* 2005, **330:**991.
9. Bürkle A, Caselli G, Franceschi C, Mariani E, Sansoni P, Santoni A, Vecchio G, Witkowski JM, Caruso C: **Pathophysiology of ageing, longevity and age related diseases.** *Immun Ageing* 2007, **4:**4.
10. Willcox BJ, Willcox DC, Todoriki H, Fujiyoshi A, Yano K, He Q, Curb JD, Suzuki M: **Caloric restriction, the traditional Okinawan diet, and healthy aging. The diet of the world's longest-lived people and its potential impact on morbidity and life span.** *Ann N Y Acad Sci* 2007, **1114:**434–455.
11. Finch CE, Crimmins EM: **Inflammatory exposure and historical changes in human life-spans.** *Science* 2004, **305:**1736–1739.
12. Licastro F, Candore G, Lio D, Porcellini E, Colonna-Romano G, Franceschi C, Caruso C: **Innate immunity and inflammation in ageing: a key for understanding age-related diseases.** *Immun Ageing* 2005, **2:**8.
13. Vasto S, Candore G, Balistreri CR, Caruso M, Colonna-Romano G, Grimaldi MP, Listi F, Nuzzo D, Lio D, Caruso C: **Inflammatory networks in ageing, age-related diseases and longevity.** *Mech Ageing Dev* 2007, **128:**83–91.
14. Candore G, Caruso C, Colonna-Romano G: **Inflammation, genetic background and longevity.** *Biogerontology* 2010, **11:**565–573.
15. Christensen K, McGue M, Petersen I, Jeune B, Vaupel JW: **Exceptional longevity does not result in excessive levels of disability.** *Proc Natl Acad Sci USA* 2008, **105:**13274–13279.
16. Farrelly C: **'Positive biology' as a new paradigm for the medical sciences. Focusing on people who live long, happy, healthy lives might hold the key to improving human well-being.** *EMBO Rep* 2012, **13:**186–188.
17. Franceschi C, Motta L, Motta M, Malaguarnera M, Capri M, Vasto S, Candore G, Caruso C: **IMUSCE: The extreme longevity: the state of the art in Italy.** *Exp Gerontol* 2008, **43:**45–52.
18. Capri M, Salvioli S, Monti D, Caruso C, Candore G, Vasto S, Olivieri F, Marchegiani F, Sansoni P, Baggio G, Mari D, Passarino G, De Benedictis G, Franceschi C: **Human longevity within an evolutionary perspective: the peculiar paradigm of a post-reproductive genetics.** *Exp Gerontol* 2008, **43:**53–60.
19. Jirillo E, Candore G, Magrone T, Caruso C: **A scientific approach to anti-ageing therapies: state of the art.** *Curr Pharm Des* 2008, **14:**2637–2642.
20. Candore G, Caruso C, Jirillo E, Magrone T, Vasto S: **Low grade inflammation as a common pathogenetic denominator in age-related diseases: novel drug targets for anti-ageing strategies and successful ageing achievement.** *Curr Pharm Des* 2010, **16:**584–596.
21. Herskind AM, McGue M, Holm NV, Sørensen TI, Harvald B, Vaupel JW: **The heritability of human longevity: a population-based study of 2872 Danish twin pairs born 1870–1900.** *Hum Genet* 1996, **97:**319–323.
22. Ljungquist B, Berg S, Lanke J, McClearn GE, Pedersen NL: **The effect of genetic factors for longevity: a comparison of identical and fraternal twins in the Swedish Twin Registry.** *J Gerontol A Biol Sci Med Sci* 1998, **53:** M441–M446.
23. Skytthe A, Pedersen NL, Kaprio J, Stazi MA, Hjelmborg JV, Iachine I, Vaupel JW, Christensen K: **Longevity studies in GenomEUtwin.** *Twin Res* 2003, **6** (5):448–454.
24. Hjelmborg J, Iachine I, Skytthe A, Vaupel JW, McGue M, Koskenvuo M, Kaprio J, Pedersen NL, Christensen K: **Genetic influence on human lifespan and longevity.** *Hum Genet* 2006, **119**(3):312–321.

25. Vaupel JW, Carey JR, Christensen K, Johnson TE, Yashin AI, Holm NV, Iachine IA, Kannisto V, Khazaeli AA, Liedo P, Longo VD, Zeng Y, Manton KG, Curtsinger JW: **Biodemographic trajectories of longevity.** *Science* 1998, **280**:855–860.
26. Terry DF, Wilcox MA, McCormick MA, Pennington JY, Schoenhofen EA, Andersen SL, Perls TT: **Lower all-cause, cardiovascular, and cancer mortality in centenarians' offspring.** *J Am Geriatr Soc* 2004, **52**:2074–2076.
27. Terry DF, McCormick M, Andersen S, Pennington J, Schoenhofen E, Palaima E, Bausero M, Ogawa K, Perls TT, Asea A: **Cardiovascular disease delay in centenarian offspring: role of heat shock proteins.** *Ann N Y Acad Sci* 2004, **1019**:502–505.
28. Hekimi S: **How genetic analysis tests theories of animal aging.** *Nat Genet* 2006, **38**:985–991.
29. Farrelly C: **Has the time come to take on time itself?.** *BMJ* 2008, **337**:147–148.
30. Olshansky SJ, Perry D, Miller RA, Butler RN: **Pursuing the longevity dividend: scientific goals for an aging world.** *Ann N Y Acad Sci* 2007, **1114**:11–13.
31. Terry DF, Sebastiani P, Andersen SL, Perls TT: **Disentangling the roles of disability and morbidity in survival to exceptional old age.** *Arch Intern Med* 2008, **168**:277–283.
32. Bonafè M, Barbi C, Storci G, Salvioli S, Capri M, Olivieri F, Valensin S, Monti D, Gonos ES, De Benedictis G, Franceschi C: **What studies on human longevity tell us about the risk for cancer in the oldest old: data and hypotheses on the genetics and immunology of centenarians.** *Exp Gerontol* 2002, **37**:1263–1271.
33. Slagboom PE, Droog S, Boomsma DI: **Genetic determination of telomere size in humans: a twin study of three age groups.** *Am J Hum Genet* 1994, **55**:876–882.
34. Beekman M, Nederstigt C, Suchiman HE, Kremer D, van der Breggen R, Lakenberg N, Alemayehu WG, de Craen AJ, Westendorp RG, Boomsma DI, de Geus EJ, Houwing-Duistermaat JJ, Heijmans BT, Slagboom PE: **Genome-wide association study (GWAS)-identified disease risk alleles do not compromise human longevity.** *Proc Natl Acad Sci USA* 2010, **107**:18046–18049.
35. Bergman A, Atzmon G, Ye K, MacCarthy T, Barzilai N: **Buffering mechanisms in aging: a systems approach toward uncovering the genetic component of aging.** *PLoS Comput Biol* 2007, **3**:e170.
36. Novelli V, Viviani Anselmi C, Roncarati R, Guffanti G, Malovini A, Piluso G, Puca AA: **Lack of replication of genetic associations with human longevity.** *Biogerontology* 2008, **9**:85–92.
37. Bulati M, Buffa S, Candore G, Caruso C, Dunn-Walters DK, Pellicanò M, Wu YC, Colonna Romano G: **B cells and immunosenescence: a focus on IgG +IgD-CD27- (DN) B cells in aged humans.** *Ageing Res Rev* 2011, **10**:274–284.
38. Colonna-Romano G, Buffa S, Bulati M, Candore G, Lio D, Pellicanò M, Vasto S, Caruso C: **B cells compartment in centenarian offspring and old people.** *Curr Pharm Des* 2010, **16**:604–608.
39. Fraser GE: **Diet as primordial prevention in Seventh-Day Adventists.** *Prev Med* 1999, **29**:S18–S23.
40. Willett W: **Lessons from dietary studies in Adventists and questions for the future.** *Am J Clin Nutr* 2003, **78**:539S–543S.
41. Rizzo NS, Sabaté J, Jaceldo-Siegl K, Fraser GE: **Vegetarian dietary patterns are associated with a lower risk of metabolic syndrome: the adventist health study 2.** *Diabetes Care* 2011, **34**:1225–1227.

doi:10.1186/1742-4933-9-5
Cite this article as: Caruso *et al.*: **"Positive biology": the centenarian lesson.** *Immunity & Ageing* 2012 **9**:5.

Montesanto et al. Immunity & Ageing 2012, **9**:6
http://www.immunityageing.com/content/9/1/6

REVIEW

Epidemiological, genetic and epigenetic aspects of the research on healthy ageing and longevity

Alberto Montesanto[†], Serena Dato[†], Dina Bellizzi, Giuseppina Rose and Giuseppe Passarino[*]

Abstract

Healthy ageing and longevity in humans result from a number of factors, including genetic background, favorable environmental and social factors and chance.

In this article we aimed to overview the research on the biological basis of human healthy ageing and longevity, discussing the role of epidemiological, genetic and epigenetic factors in the variation of quality of ageing and lifespan, including the most promising candidate genes investigated so far. Moreover, we reported the methodologies applied for their identification, discussing advantages and disadvantages of the different approaches and possible solutions that can be taken to overcome them. Finally, we illustrated the recent approaches to define healthy ageing and underlined the role that the emerging field of epigenetics is gaining in the search for the determinants of healthy ageing and longevity.

Keywords: Ageing, Longevity, Genetic variation, Epigenetic modifications

Background

The past few decades witnessed a growing social and scientific interest in studies on human ageing and longevity. This interest is primarily due to the social burden connected to the extraordinary increase of the elder population in developed countries, which implies an increase of the subjects which are not autonomous and are affected by invalidating pathologies [1,2]. In Italy, for instance, in 1961 the population aged 65 and older was 4.8 million (9.5% of the total population), while in 1981 this number increased up to 7.5 million (13.2% of the total population) and in 2011 it grew up to 12.3 million (20.3% of the total population). In addition, the population aged 90 and older is growing at a faster pace as it has triplicated in the last 20 years (data from population Census and from http://www.istat.it). Proportionally, life expectancy at birth increased from a medium value of 44 years (44.2 for males and 43.7 for females) in 1905 to more than 80 years (79.4 for males and 84.5 for females) in 2011. Similar figures are reported for all developed countries, while in developing countries life expectancy grows very fast as soon as infant mortality is reduced,

with the exception of some areas, namely in Africa, where AIDS infection dramatically affects life expectancy of adults [2].

Epidemiological evidence for a genetic component to variation in human lifespan comes from twin studies and family studies. By comparing life span in twins, researchers have found that approximately 25% of the overall variation in human lifespan can be attributed to genetic factors [3-5], which become more relevant for extreme longevity [6]. Conditioning factors, that arise in the first part of life (socio-economic state of parents, education and month of birth, which has been found to reflect the environmental conditions during the prenatal and early postnatal period), account for another 25% of such variability; life circumstances at adult and old age (including socio-economic status and medical assistance) may account for about the remaining 50% [7].

Family-based studies demonstrated that parents, siblings and offspring of long-lived subjects have a significant survival advantage when compared with the general population [8-12]. Moreover, these studies indicated that long-lived individuals and their children experienced a lower incidence of age related diseases and a higher degree of physical functioning and autonomy, when compared to appropriate selected controls [13-15]. However, how much of this reported survival advantage is due to

* Correspondence: g.passarino@unical.it
† Contributed equally
Department of Cell Biology, University of Calabria, Ponte Pietro Bucci cubo 4
C, 87036 Rende, CS, Italy

common genetic factors or to a shared environment remained unclear. By using the original approach to adopt an intra-family control group, two different studies [16,17] confirmed that a substantial contribution in the familiarity observed in the above cited works was attributable to genetic variation, so prompting the research to deeply investigate the genetic variants favoring human longevity.

In this paper we will review the literature on the studies on the genetic of human longevity and the discussions there has been on the different approaches that can be used in this field. In addition, we will report the new approaches that have been proposed to define the healthy ageing, as the correct definition of healthy ageing is the first step to understand its genetic basis. Finally, we will outline some recent advance in the epigenetic studies of ageing, as epigenetics, a bridge between genetics and environment, might explain many aspects of ageing and longevity.

Genetic variability and human longevity

The studies aimed to understand the genetic basis of longevity in humans have been carried out under the hypothesis that unfavorable genotypes should be dropped out of the population by a sort of "demographic selection" [18] which finally results in an enrichment of favorable genotypes in the gene pool of long lived people [19-21]. These studies have preliminarily faced the difficulty of clearly defining the phenotype under study. In fact, longevity is a dynamic phenomenon, where the definition changes in relation to the individual birth cohort. Indeed, survival curves change with time, in relation to the birth year of the cohort, thus medium age at death progressively increases with time modifying the number of subjects who can be defined as "long lived" [7,22]. In this frame, demographic analyses allowed to show that around the age of 90 years there is a clear deceleration of the age related mortality rate [23], suggesting that the subjects surviving to this age might be considered the long lived subjects who have survived the "demographic selection" mentioned above [24].

To date, many approaches have been adopted in order to disentangle the genetic from the environmental effects on human longevity, ranging from different sample design to data analysis approaches [25]. Among the different sampling strategies adopted in the field of human longevity research, a first distinction should be made between family-based and population-based studies.

Family-based studies

At family level, the ASP design represents the typical non-parametric strategy allowing both linkage and association to be tested [26]. At population level, cross-sectional (or case-control) cohort (longitudinal or follow-up) and case-only studies represent the most common design strategies

providing important insights into the genetics of human longevity. Family-based designs show the unique advantages over population-based designs, as they are robust against population admixture and stratification. On the other hand, it is evident the difficulty to collect enough families, especially for late-onset complex traits such as lifespan, in which parental genotype information is usually missing. Despite these problems, non-parametric linkage analysis was attempted to localize genes implicated in human longevity. One of the first attempts to identify genetic regions co-segregating with the longevity phenotype by using an ASP approach has been carried out by Puca and co-workers [27]. Scanning the whole genome by applying non-parametric linkage analysis to long-lived sib-pairs from USA they reported a region on chromosome 4 that could possibly harbor a gene affecting human longevity. In a following association-based fine-mapping experiment of the region, *MTTP* was identified as the gene most probably responsible for the observed linkage peak reported [28]. However, the association observed in this sample could not be replicated neither in a larger French sample of long-lived individuals nor in a sample of German nonagenarians and centenarians [28,29].

Among the studies using an ASP approach it is worth noticing the original study design adopted in the ECHA project [30]. The authors, by using cousin-pairs born from siblings who were concordant or discordant for the longevity trait, analyzed two chromosomal regions already known to encompass longevity-related genes. Although no significant differences emerged between the two groups of cousin-pairs (probably due to insufficient sample size) this study provided important insights to better dimension future sampling campaigns to study-genetic basis of human longevity. In particular the GEHA project [31] was launched in 2004 and was aimed to the sampling of an unprecedented number (2500) of nonagenarians sib-pairs from all over the Europe, to be analyzed for selected chromosomal regions previous related to the longevity trait, and for discovering new regions by a whole genome approach. Behind the scientific results still to be published, GEHA clearly represents an example of standard recruitment methodology, both for collecting biological samples and phenotypic information by home-based questionnaires, the last very crucial for the definition of phenotype [31].

Population case-control studies

Population case-control studies comparing long lived samples with younger controls of the same population may provide a powerful and more efficient alternative, especially when associated to the recent advances in genomic and statistical techniques. They are more powerful than family designs for detecting genes with low effect and gene-gene interactions [32]. However, these cross-sectional studies may suffer from the lack of appropriate control groups, as cohort specific effects may confound comparisons between very old people (for example centenarians) and younger cohorts [33]. The problem is hindered by the rapid changes of human societies that increase the level of population heterogeneity, thus introducing a further complicating factor. To cope with these problems, algorithms which integrate genetic and demographic data have been proposed [22,24,34,35]. Genetic-demographic methods allow the estimation of hazard rates and survival functions in relation to candidate alleles and genotypes. In such a way it is possible to compare survival functions between individuals carrying or not carrying a candidate allele or genotype without introducing arbitrary age classes, and taking into account cohort effects in mortality changes. Furthermore, the addition of demographic to genetic data not only is able to reveal sex- and age-specific allelic effects, but also permit a rational definition of the age classes to be screened [24]. Moreover, from the application of genetic-demographic model to longevity association studies, it emerged that genetic factors influence human survival in a sex- and age-specific way. In fact, in agreement with demographic data, genetic variability plays a stronger role in males than in females and in both genders its impact is especially important at very old ages [6,17,24].

Multi-locus approaches

Most gene-longevity association studies have focused on a single or a few candidate genes. However, common genetic variants with important effects on human longevity are unlikely to exist because of the rather low genetic contribution to the trait. In addition, given the complexity of the trait, the main effects of the individual loci may be small or absent, while multiple genes with a small effect may interact in an additive manner and affect survival at old ages. In such a case, a single-locus approach may not be suitable, failing in finding positive results of associations. Thus, given the technical improvement of typing techniques, multi-locus association approaches which takes into account epistatic interactions among different genes, have become of age [36].

These approaches represent specific and important statistical challenges. A flexible framework to tackle these challenges and for modeling the relationship between multiple risk loci and a complex trait makes use of logistic regression techniques [24,37]. Since from a statistical point of view epistasis corresponds to an interaction between genotypes at two or more loci, the same regression techniques have been easily extended to the analysis of gene-gene and gene-environment interactions in complex phenotypes, both at genome-wide and smaller scale studies level [38,39].

In some studies, different loci clustered in haplotypes are analyzed. In general haplotype-based association analysis brings new possibilities and difficulties. They exhibit more power than single-marker analysis for genetic association studies since they incorporate linkage disequilibrium information [40-42]. Conversely, the main difficulty is that haplotypes are often not directly observable, especially for late-onset complex traits such as lifespan, owing to phase uncertainty. Methods based upon likelihood can be extended to deal with kind of problem, most conveniently by use of the EM algorithm. Among these, the score tests proposed by Schaid et al. [43] are the most popular. Among the methods developed for haplotype-based multi-locus analysis of human survival, the original studies carried out by Tan et al. involving both cross-sectional [44] and cohort [45] designs studies of unrelated individuals are worth noticing.

Further improvements in high-throughput technology, associated to the recent advances in genomic knowledge, have made whole genome genotyping (> 100,000 SNPs) more accessible. Indeed, GWAS are at present widely used to find genetic variants contributing to variation in human lifespan [27,46-52]. In particular, Sebastiani and co-workers, consistently with the hypothesis that the genetic contribution is largest at the oldest ages and that long-lived individuals are endowed with multiple genetic variants with a single small effect, undertook a genome-wide association study of exceptional longevity, building a genetic profile including 281 SNPs able to discriminate between 800 centenarians cases and 900 healthy controls. The "genetic signatures of exceptional longevity" and relative subject-specific genetic risk profile which were obtained can provide important insights to dissect the unique complex phenotype into sub-phenotypes of exceptional longevity.

From a statistical point of view, the analysis of GWAS data presents several statistical challenges including data reduction, interaction of variables and multiple testing. Although these challenges are new to statistics, the magnitude of the present datasets is unprecedented.

After all these considerations, the most reasonable approach, for taking into account a great number of single polymorphisms spread along the genome without losing the biological relevance of candidate genes in biochemical pathways, which may be reasonably related to the trait, seems to be to use a candidate regions approach combined with a minimal number of "tagging" SNPs, efficiently capturing all the common genetic variation in the assayed genomic region [24,53-56]. This hybrid tagging-functional approach, by selecting the maximally informative set of tag SNPs in candidate-gene/candidate region for an association study, promises to shed a light in the genetic determinants of complex traits in general, and hopefully in human longevity too [57].

Candidate genes and candidate pathways in human longevity

By using the approaches described above, many candidate genes have been investigated to identify alleles that are either positively or negatively selected in the centenarian population as consequence of a demographic pressure. For many years, genetic analyses were focused on single genetic variants, by using the classical "candidate gene" approach. Candidates were found among human orthologous of experimental model genes, where the existence of specific mutations (*age-1, daf2, sir2, methuselah, p66*) able to extend or reduce lifespan has been reported [58-62]. In laboratory models, all the longevity genes identified have primary roles in physiological processes and especially in signal transduction; therefore it seems that natural selection does not select for genes that cause ageing in these organisms, but rather ageing occurs as a result of pleiotropic effects of genes that specify other fundamental processes.

In providing these insights, invertebrate studies motivated a lot the search for human genes involved in longevity and provided candidate genes, sometime successfully found associated with human longevity too (i.e. *KLOTHO, FOXO3a, SIRT3; UCPs;* [20,63-66]. However, these studies revealed also many challenges and claimed for caution that should be used when investigating human candidate genes identified by their orthology in animal models [33]. Another important category of candidate genes for ageing research are those involved in age-related diseases (in particular, cardiovascular diseases, Alzheimer's, cancer and auto-immune diseases) and genes involved in genome maintenance and repair (in particular, those involved in premature ageing syndromes such as Werner syndrome). The underlying hypothesis is that long lived persons should not present in their DNA any risk factors involved in pathologies. On the contrary, long lived individuals harbor genetic risk factors for age-related diseases [67,68] as recently underlined also by GWAS data, reporting as very long lived individuals share the same number of risk alleles for coronary artery disease, cancer, and type 2 diabetes than younger controls from the same population, thus suggesting that human longevity is not compromised by the cumulative effect of a set of risk alleles for common disease [69]. These studies support the existence of buffering mechanisms operating in the determination of human longevity, probably through the presence of favorable genotypes contrasting the deleterious effect of age-related disease genes: as a result, the frequency of deleterious genotypes may increase among individuals with extreme lifespan because their protective genotype allows disease-related genes to accumulate [70].

Recently, from the study of a single gene and starting again from the evidences in experimental models, which suggest the existence of evolutionary conserved networks that regulates lifespan and affects longevity across species, research moved into the study of whole metabolic pathways, where to find candidate genes for human longevity. From worms (*C. elegans*), to fruit flies (*Drosophila*), and mammals (*mouse*), pathways related to the regulation of energy homeostasis, cell maintenance, nutritional sensing, stress response signalling to internal or external environmental insults, by an efficient non inflammatory response, and DNA repair/maintenance have been shown to critically modulate lifespan [62,71] so harboring interesting candidate genes for longevity research. The insulin/IGF-1 pathway and downstream effectors, such as FOXO, are among of the most promising in this sense. Mutations affecting this pathway show effects on longevity from invertebrates to mammals, with several longevity mutants altering key components of the pathway, as for example the increased lifespan of mice heterozygous for the IGF1 receptor knockout1 [72]. Moreover, the downstream transcription factor DAF-16 (FOXO) regulates the expression of several genes involved in stress resistance, innate immunity, metabolic processes and toxin degradation [73]. Other interesting pathways for human longevity are represented by the TOR signalling, a major nutrient-sensing pathway, whose genetic down-regulation can improve health and extend lifespan in evolutionarily distant organisms such as yeast and mammals [74] and the recently deep investigated UCP pathway, a family of inner mitochondrial membrane proteins responsible for uncoupling substrate oxidation from ATP synthesis, whose expression was demonstrated to affect lifespan from fruit flies to mouse, somehow mimicking the metabolic and lifespan effects of caloric restriction (see [65] and references therein).

In humans, the most relevant results found by association studies in long lived cohorts, identified genes involved in GH/IGF-1/Insulin signaling (*GHR, IGF1R, FOXO3A*), antioxidant (*SOD1, SOD2, PON1, FOXO3A*), inflammatory (*IL6, CETP, Klotho*) pathways, silencing (*SIRT1 and SIRT3*), elements of lipid metabolism (*APOE, APOB, ACE, APOC3, MTTP*) and stress resistance (*HSPA1A and HSPA1L*) [[19,33,75-81] and references therein]. However, most of these results, with the exception of *APOE* and *FOXO3A*, were not reproduced in some of the replication studies [29,82], probably because of problems in study design and publication bias. This points to the need of larger populations for case-control studies in extreme longevity, use of replication cohorts from different populations and appropriate multiple comparisons tests to reduce the bias of these kind of studies [83].

Functional consequences of genetic variants associated with human longevity

Coupled with the rapid advances in high-throughput sequencing technologies, it is now feasible to comprehensively analyze all possible sequence variants in candidate genes segregating with a longevity phenotype and to investigate the functional consequences of the associated variants. A better understanding of the functional genes that affect healthy longevity in humans may lead to a rational basis for intervention strategies that can delay or prevent age-related diseases. Genome-wide expression profiles in different tissues reported changes of gene expression occurring with age. In this sense, two main works deserve attention. Kerber and collaborators, who analyzed the gene expression profiles of 2151 house-keeping genes in cultured cell lines from 104 adults belonging to 31 Utah families, aged 57-97 years, searching for stable variation in gene expressions that affect or mark longevity. They found different genes exhibiting associations with either mortality or survival [84], 10% decreased in expression with age, and 6% increased with age. Significant association both with age and survival was observed for CDC42, belonging to the DNA repair pathway and CORO1A, an actin-binding protein with potentially important functions in both T-cell mediated immunity and mitochondrial apoptosis [85], underlying the potential importance of these metabolic pathway in longevity determination. More recently, Slagboom and co-workers [81] compared the expression profiles of candidate genes in a limited number of subjects (50 for each group) among offspring of long-lived subjects and their partners. Among the differentially expressed genes, they observed a decreased expression of genes in the mTOR pathway in the members of long-lived families. Although it is likely that epigenetic factors may also play a large role [86] and the results should be replicate in a larger sample, it is clear that by combining the molecular epidemiological studies with a genomic approach may provide a step further towards the identification of early and possibly causal contributions to the ageing and human longevity process.

The special case of the mitochondrial genome

Human ageing is characterized by a gradual reduction in the ability to coordinate cellular energy expenditure and storage (crucial to maintain energy homeostasis), and by a gradual decrease in the in the ability to mount a successful stress response [87]. These physiological changes are typically associated with changes in body composition (i.e. increase in fat mass and the decline in fat-free mass), and with a chronic state of oxidative stress with important consequences on health status [88]. Mitochondrial function is crucial in these processes, being mitochondria the main cellular sites controlling energy metabolism and the redox state. Mitochondria are considered as key components of the ageing process, playing a pivotal role in cell survival and death since they contribute to many cellular functions, including bioenergetics, protection from oxidative damage, maintenance of mtDNA and cell death

[89]. Moreover, in addition to ATP production, mitochondria form a complex metabolic network which is crucially involved in glucose sensing/insulin regulation, intracellular $Ca2+$ homeostasis and many metabolic signaling pathways [90]. On the other hand, mitochondria are the major producers of ROS and at the same time targets of ROS toxicity. Consequently, the maintenance of a healthy mitochondria population represents a major target of a well functioning organism, for preserving many physiological functions, such as neurotransmission [91]. Starting from the important role of this organelle in the cell homeostasis, the effect of both inherited and somatic variability of mtDNA in ageing and longevity has been deeply investigated, resulting complex and sometime controversial [92].

An accumulation of mtDNA somatic mutations occurs with age, and many studies have reported an association between mtDNA mutations and ageing, particularly in post-mitotic neuronal cells [93]. A number of mutations not associated to diseases have been fixed along time in the mtDNA sequence, to form a series of population-specific lineages that can be identified by the presence of conserved groups of haplotypes (haplogroups). These germline inherited mtDNA variants (haplogroups and their subclassification into subhaplogroups on the basis of specific mutations identified by sequence analysis of the D-loop region) are used for tracing back the origin of populations or in forensic analyses [94]. Considered biochemically neutral, the mtDNA inherited variability is probably able to differently modulate the mitochondrial metabolism [95]. mtDNA haplogroups have been positively associated with mitochondrial, complex diseases and ageing [96,97]. In particular, in Caucasians the haplogroup J is over-represented in long-living people and centenarians, thus suggesting a role for this mtDNA variant in longevity [98]. As for somatic variations, tissue specific mutations occurring in the mtDNA control region have been proposed to provide a survival advantage, i.e. the C150T transition [99]. Data analyzing the occurrence and accumulation of C150T mutation in centenarians' relatives and long lived sib pairs demonstrated a genetic control on the mtDNA heteroplasmy (i.e. the presence of different molecule of mutant/wild type mtDNA), suggesting the existence of nuclear genetic factor influencing their accumulation [100,101]. The observation that the nuclear genome contributes to mtDNA heteroplasmy marks the importance of the mitochondrial-nucleus cross-talk in modulating mitochondrial function and cellular homeostasis and, consequently, quality of ageing and lifespan [102]. Such a nuclear-mitochondrial cross-talk was firstly observed in yeast, where a compensatory mechanism, named "retrograde response" has been described, allowing to mutant strains of yeast to cope with mtDNA impairments by up-regulating the expression of stress-responder nuclear genes [103] and leading to a significant increased lifespan.

The first experimental evidence that a similar mechanism has been maintained in higher organisms, including humans, comes from cytoplasmic hybrid or cybrid experiments (i.e. cell lines differ only in the source of their mtDNA), where it was found that cells characterized by different mtDNA haplogroups, differently expressed stress responder nuclear genes [104,105], thus suggesting that the retrograde response mechanism may represent an evolutionary conserved strategy for the age-related remodeling of organismal functions.

On the whole, although the involvement of mtDNA variability in ageing and longevity is undisputed, the role of mtDNA and its mutations, either inherited or somatically acquired, in human longevity is far from being clear. The use of high-throughput technologies and the extensive analysis, possibly at the single cell level, of different tissues and cell types derived from the same individual will help in disentangling the complexity of mtDNA in ageing and longevity.

The maintenance of telomere length

Genomic instability has been widely recognized as a crucial mechanism in both ageing and age-related diseases. The progressive shortening of telomeres, probably the most important marker of chromosome integrity, is associated with increased risk of several age-related disease comprised cancer and mortality [106,107]. Telomeres play a central role in maintaining the chromosome stability, preventing the inappropriate activation of DNA damage pathways, and regulating cell viability, by triggering signals of ageing to normal cells to senesce when telomeres stop their functioning [108]. Their length is controlled by telomerase. In normal human cells telomerase is expressed in stem cells, cells that need to actively divide (like immune cells) and is barely, or not expressed at all in differentiated somatic cells. However, higher expression of telomerase strongly correlated with carcinogenesis, with approximately 85%-90% of human cancers showing higher enzymatic activity [109]. Furthermore, suppression of telomerase activity in telomerase-positive cancer cells results in cell death and tumor growth inhibition [110], highlighting the critical role of telomerase in facilitating and enabling cancer cell proliferation. On the contrary, high telomere stability correlates with human longevity, with healthy individuals showing significantly longer telomeres than their unhealthier counterparts [68,111]. Longer telomeres are associated with protection from age-related diseases, better cognitive function and lipid profiles, thus may confer exceptional longevity [112]. The understanding of the complex tradeoff between cancer development and long life in relation to telomere maintenance represents one of the most intriguing challenges for researchers in human longevity. Considering these evidences,

centenarians may represent the best example of a well preserved telomere length, harboring the right compromise of having longer telomeres and never have been affected by cancer or survived to a cancer episode, so may represent optimal control population for association studies aimed to disentangle the complex role of telomere maintenance in age-related diseases and ageing.

Successful ageing and frailty

Although ageing is a general phenomenon, it is clear that a great inter-individual variability on the rate and quality of ageing can be observed [33]. Following the paradigm "Centenarians as a model for healthy ageing", centenarian studies have allowed to identify a number of characteristics associated with extreme longevity. For example, nonagenarian and centenarian men are generally taller and heavier than women of corresponding age and have a greater amount of muscle and trunk fat, whereas women are small and show a marked peripheral adipose distribution [113]. Furthermore, food preferences, marital status, personality and coping strategies, levels of family support, and education have all been linked with successful late-life ageing [113-118]. However, whether centenarians represent healthy ageing still remains an open question. Franceschi and co-workers recognized that on the basis of their functional status centenarians might be classified into three categories [119]. Most of them suffer of disabilities or diseases [120], and in general they experience a loss of independence [1], but a minority of them are still in quite a good health. According to this perspective, centenarians are not the most robust subjects of their age cohort, but rather those who better adapted and re-adapted from both biological and non-biological point of view, and in general they constitute a very heterogeneous group of individuals [119]. Hence several studies searched for indicators of health and functional status in old and very old subjects by which objective phenotypes could be defined [121-126]. From these studies the concept of frailty emerged as a distinct clinical entity, characterized by a state of vulnerability for adverse health outcomes, such as hospitalization or death, and therefore correlated to co-morbidity, disability and increased mortality hazard [127]. The "frailty" syndrome of the elderly is mainly correlated to the decline of homeostatic capacity of the organism, which implies the decline of different physiological systems, such as the neuromuscular and the cognitive systems, and which leads to a significant increase of disability, comorbidity and death risk [121]. The frailty declines with age and make less efficient the metabolic pathways for the conservation, the mobilization and the use of the nutrients, thus representing the physiologic precursor and etiologic factor in disability, due to its central features of weakness, decreased endurance, and slowed performance [121]. Therefore, the identification of a precise frailty phenotype could help to recognize homogeneous population groups enriched of genetic risk factors predisposing to a poor quality of ageing. How to measure frailty? First of all, because population specificity was demonstrated in the quality of ageing [128], it is necessary to carry out population specific surveys to define the tools which are able to highlight within each population groups of subjects with homogeneous "ageing phenotype". Among the methodologies used to classify homogenous subgroups within each population, the cluster analysis proved to be very useful to identify groups of subjects homogeneous with respect to chosen variables. As for the parameters to be used for the classification, cognitive, psychological and functional measures turned out to be the most effective to identify the frailty phenotype, since these parameters condense most of the frailty cycle that occurs in the elderly [122]. In particular, classification variables useful for grouping individuals respect their frailty status are represented by SHRS, ADL, HG strength and MMSE [129,130]. This kind of classification, which allows to define three main frailty groups (i.e. frail, pre-frail and non frail subjects), was firstly applied to a Southern Italian population, and proved to be able to foresee health status by the analysis of perspective survival. In particular, a longitudinal study showed a differential incidence of mortality after 18 and 36 months follow-up of the different groups identified [129]. The proposed classification was replicated in two large longitudinal Danish samples [130], where different ageing conditions had been previously described [128], confirming the predictive soundness after a 10-years follow up. In addition, in the same work the differential effect of distinct parameters on survival was estimated, founding that high values of HG and MMSE induced a higher probability of surviving, while being male, having a low ADL or a poor SRHS tended to reduce expected survival time. Furthermore, the presence of a genetic influence on frailty variance was suggested by the estimation of heritability of the frailty status, where it was found that the additive genetic component accounts for 43% of the overall variability of frailty levels between couple of twins. In line with previous findings, the estimate was higher in males than in females, consistent with the hypothesis that frailty status of men is more related to the genetic background while the frailty conditions of females are more dependent on environmental factors. In addition, as for lifespan, the influence of the genetic component on frailty status was found higher at advanced ages.

On the whole, this approach, which is based on population-specific data under study and does not use any a priori thresholds, may be very promising for an objective identification of frail subject. This may be a very important task for future societies, helping to address specific medical care, by tailoring treatments on the basis of the real needs of each single patient, especially of pre-frail and frail older patients with multiple chronic conditions and reduced life expectancy, finally preventing the effects of frailty.

The role of epigenetics in human ageing and longevity

Epigenetic modifications indicate the sum of heritable changes, such as DNA methylation, histone modification and miRNA expression, that affect gene expression without changing the DNA sequence [131]. It is becoming clear that epigenetic information is only partially stable and destined to change across the lifespan representing a drawbridge between genetics and environment. Epigenetic variations have been suggested to have an important role in cellular senescence, tumorigenesis and in several diseases including type-2 diabetes, cardiovascular and autoimmune diseases, obesity and Alzheimer disease [132]. A correlation between epigenetic DNA modifications and human lifespan has been shown by Fraga et al. [133], who found that global and locus-specific differences in DNA methylation in identical twins of different ages are influenced by environmental factors and lifestyle. Most studies demonstrated that ageing is associated with a relaxation of epigenetic control; from one side, a decrease in global cytosine methylation has been found during ageing both in vivo and in vitro studies, mostly due to the demethylation in transposable repetitive elements [134,135]. On the other hand, an age-related hypermethylation has been observed in promoter regions of specific genes, such as those genes involved in cell cycle regulation, tumor-cell invasion, apoptosis, metabolism, cell signaling and DNA repair, with a consequent decrease of correspondent mRNA levels, confirming the potential role of these pathways in human ageing [136-143]. Moreover, recent studies reported as different epigenetic profiles can be associated with a different quality of ageing. Bellizzi and co-workers [144], studying the distribution of methylation pattern in a sample of elderly subjects stratified according to their quality of ageing (described by their scores in specific functional, cognitive and psychological tests), found that the level of methylation is correlated with the health status in the elderly. In particular, a significant decrease in the global DNA methylation levels was associated with functional decline, suggesting that the relaxation of the epigenetic control in ageing is specifically associated with the functional decline rather than with the chronological age of individuals. These results confirm that epigenetic variations, which in turn depend on hereditary, environmental and stochastic factors, may play an important role in determining physiological changes associated to old age.

Conclusions

Despite the enormous technical progresses, which allow to analyzed many single variants as well as the coordinated expression of many genes together by high-throughput platforms, many challenges still remain to be faced by the researchers trying to identify genetic and non genetic variants associated with human longevity. A close partnership between gerontologists, epidemiologists, and geneticists is needed to take full advantage of emerging genome information and technology and bring about a new age for biological ageing research. In addition, we believe that the next future will see much progresses in our understanding of the longevity trait, mainly coming from the integration of genetics and epigenetics information by multidisciplinary approaches, to the aim of obtaining an overall picture of what successful ageing is.

Abbreviations

ACE: Angiotensin I converting enzyme; ADL: Activity of Daily Living; APOE/B: Apolipoprotein E/B; APOC3: Apolipoprotein C-III; ASP: Affected Sib-Pairs; ATP: Adenosine triphosphate; CDC42: Cell division cycle 42; CETP: Cholesteryl ester transfer protein; CORO1A: Coronin, actin binding protein, 1A; DNA: Deoxyribonucleic Acid; ECHA: European Challenge for Healthy Aging; EM: Maximum Estimation; FOXO3A: Forkhead box O3; GEHA: Genetics of Healthy Aging; GH: Growth Hormone; GHR: Growth hormone receptor; GWAS: Genome-Wide Association Studies; HG: Hand grip; HSPA1A: Heat shock 70 kDa protein 1A; HSPA1L: Heat shock 70 kDa protein 1-like; IGF-1: Insulin Growth Factor 1; IGF1R: Insulin-like growth factor 1 receptor; IL6: Interleukine 6; miRNA: MicroRNA; MMSE: Mini Mental State Examination; mRNA: Messenger RNA; mtDNA: Mitochondrial DNA; mTOR: Mitochondrial Target Of Rapamycin; MTTP: Microsomal Triglyceride Transfer Protein; PON1: Paraoxonase 1; ROS: Reactive oxygen species; SOD1: Superoxide dismutase 1, soluble; SHRS: Self-reported health status; SIRT1/3: SIR2-like protein 1/3; SNP: Single Nucleotide Polymorphism; SOD: Superoxide dismutase; TOR: Target Of Rapamycin; UCP: Uncoupling Protein.

Authors' contributions

AM and SD wrote the first draft; subsequent drafts were written by DB, GR and GP who also supervised the review process; all the authors edited the manuscript and approved its final version.

Competing interests

The authors declare that they have no competing interests.

Received: 3 April 2012 Accepted: 23 April 2012 Published: 23 April 2012

References

1. Christensen K, McGue M, Petersen I, Jeune B, Vaupel JW: Exceptional longevity does not result in excessive levels of disability. *Proc Natl Acad Sci USA* 2008, **105**(36):13274-13279.
2. United Nations, Department of Economic and Social Affairs, Population Division: World Population Prospects: The 2010 Revision, Highlights and Advance Tables. 2011, ESA/P/WP.220.
3. Herskind AM, McGue M, Holm NV, Sørensen TI, Harvald B, Vaupel JW: The heritability of human longevity: a population-based study of 2872 Danish twin pairs born 1870-1900. *Hum Genet* 1996, **97**(3):319-323.
4. Ljungquist B, Berg S, Lanke J, McClearn GE, Pedersen NL: The effect of genetic factors for longevity: a comparison of identical and fraternal twins in the Swedish Twin Registry. *J Gerontol A Biol Sci Med Sci* 1998, **53**(6):M441-M446.
5. Skytthe A, Pedersen NL, Kaprio J, Stazi MA, Hjelmborg JV, Iachine I, Vaupel JW, Christensen K: Longevity studies in GenomEUtwin. *Twin Res* 2003, **6**(5):448-454.
6. Hjelmborg J, Iachine I, Skytthe A, Vaupel JW, McGue M, Koskenvuo M, Kaprio J, Pedersen NL, Christensen K: Genetic influence on human lifespan and longevity. *Hum Genet* 2006, **119**(3):312-321.
7. Vaupel JW, Carey JR, Christensen K, Johnson TE, Yashin AI, Holm NV, Iachine IA, Kannisto V, Khazaeli AA, Liedo P, Longo VD, Zeng Y, Manton KG, Curtsinger JW: Biodemographic trajectories of longevity. *Science* 1998, **280**(5365):855-860.
8. Gudmundsson H, Gudbjartsson DF, Frigge M, Gulcher JR, Stefánsson K: Inheritance of human longevity in Iceland. *Eur J Hum Genet* 2000, **8**(10):743-749.
9. Kerber RA, O'Brien E, Smith KR, Cawthon RM: Familial excess longevity in Utah genealogies. *J Gerontol A Biol Sci Med Sci* 2001, **56**(3):B130-B139.
10. Perls T, Shea-Drinkwater M, Bowen-Flynn J, Ridge SB, Kang S, Joyce E, Daly M, Brewster SJ, Kunkel L, Puca AA: Exceptional familial clustering for extreme longevity in humans. *J Am Geriatr Soc* 2000, **48**(11):1483-1485.
11. Perls TT, Wilmoth J, Levenson R, Drinkwater M, Cohen M, Bogan H, Joyce E, Brewster S, Kunkel L, Puca A: Life-long sustained mortality advantage of siblings of centenarians. *Proc Natl Acad Sci USA* 2002, **99**(12):8442-8447.
12. Willcox BJ, Willcox DC, He Q, Curb JD, Suzuki M: Siblings of Okinawan centenarians share lifelong mortality advantages. *J Gerontol A Biol Sci Med Sci* 2006, **61**(4):345-354.
13. Terry DF, Wilcox M, McCormick MA, Lawler E, Perls TT: Cardiovascular advantages among the offspring of centenarians. *J Gerontol A Biol Sci Med Sci* 2003, **63**(7):706.
14. Terry DF, Wilcox MA, McCormick MA, Pennington JY, Schoenhofen EA, Andersen SL, Perls TT: Lower all-cause, cardiovascular, and cancer mortality in centenarians' offspring. *J Am Geriatr Soc* 2004, **52**:2074-2076.
15. Atzmon G, Rincon M, Schechter CB, Shuldiner AR, Lipton RB, Bergman A, Barzilai N: Lipoprotein genotype and conserved pathway for exceptional longevity in humans. *PLoS Biol* 2006, **4**(4):e113.
16. Schoenmaker M, de Craen AJ, de Meijer PH, Beekman M, Blauw GJ, Slagboom PE, Westendorp RG: Evidence of genetic enrichment for exceptional survival using a family approach: the Leiden Longevity Study. *Eur J Hum Genet* 2006, **14**(1):79-84.
17. Montesanto A, Latorre V, Giordano M, Martino C, Domma F, Passarino G: The genetic component of human longevity: analysis of the survival advantage of parents and siblings of Italian nonagenarians. *Eur J Hum Genet* 2011, **19**(8):882-886.
18. Perls T, Kunkel L, Puca A: The genetics of aging. *Curr Opin Genet Dev* 2002, **12**(3):362-369.
19. Altomare K, Greco V, Bellizzi D, Berardelli M, Dato S, DeRango F, Garasto S, Rose G, Feraco E, Mari V, Passarino G, Franceschi C, De Benedictis G: The allele (A)(-110) in the promoter region of the HSP70-1 gene is unfavorable to longevity in women. *Biogerontology* 2003, **4**(4):215-220.
20. Bellizzi D, Rose G, Cavalcante P, Covello G, Dato S, De Rango F, Greco V, Maggiolini M, Feraco E, Mari V, Franceschi C, Passarino G, De Benedictis G: A novel VNTR enhancer within the SIRT3 gene, a human homologue of SIR2, is associated with survival at oldest ages. *Genomics* 2005, **85**(2):258-263.
21. Franceschi C, Olivieri F, Marchegiani F, Cardelli M, Cavallone L, Capri M, Salvioli S, Valensin S, De Benedictis G, Di Iorio A, Caruso C, Paolisso G, Monti D: Genes involved in immune response/inflammation, IGF1/insulin pathway and response to oxidative stress play a major role in the genetics of human longevity: the lesson of centenarians. *Ageing Dev* 2005, **126**(2):351-361.
22. Dato S, Carotenuto L, De Benedictis G: Genes and longevity: a genetic-demographic approach reveals sex- and age-specific gene effects not shown by the case-control approach (APOE and HSP70.1 loci). *Biogerontology* 2007, **8**(1):31-41.
23. Yashin AI, Ukraintseva SV, De Benedictis G, Anisimov VN, Butov AA, Arbeev K, Jdanov DA, Boiko SI, Begun AS, Bonafe M, Franceschi C: Have the oldest old adults ever been frail in the past? A hypothesis that explains modern trends in survival. *J Gerontol A Biol Sci Med Sci* 2001, **56**(10): B432-B442.
24. Passarino G, Montesanto A, Dato S, Giordano S, Domma F, Mari V, Feraco E, De Benedictis G: Sex and age specificity of susceptibility genes modulating survival at old age. *Hum Hered* 2006, **62**(4):213-220.
25. Tan Q, Kruse TA, Christensen K: Design and analysis in genetic studies of human ageing and longevity. *Ageing Res Rev* 2006, **5**(4):371-387.
26. Elston RC: Segregation analysis. *Adv Hum Genet* 1981, **11**:63-120, 372-3.

27. Puca AA, Daly MJ, Brewster SJ, Matise TC, Barrett J, Shea-Drinkwater M, Kang S, Joyce E, Nicoli J, Benson E, Kunkel LM, Perls T: **A genome-wide scan for linkage to human exceptional longevity identifies a locus on chromosome 4.** *Proc Natl Acad Sci USA* 2001, **98(18)**:10505-10508.

28. Geesaman BJ, Benson E, Brewster SJ, Kunkel LM, Blanché H, Thomas G, Perls TT, Daly MJ, Puca AA: **Haplotype-based identification of a microsomal transfer protein marker associated with the human lifespan.** *Proc Natl Acad Sci USA* 2003, **100(24)**:14115-14120.

29. Nebel A, Croucher PJ, Stiegeler R, Nikolaus S, Krawczak M, Schreiber S: **No association between microsomal triglyceride transfer protein (MTP) haplotype and longevity in humans.** *Proc Natl Acad Sci USA* 2005, **102(22)**:7906-7909.

30. De Rango F, Dato S, Bellizzi D, Rose G, Marzi E, Cavallone L, Franceschi C, Skytthe A, Jeune B, Cournil A, Robine JM, Gampe J, Vaupel JW, Mari V, Feraco E, Passarino G, Novelletto A, De Benedictis G: **A novel sampling design to explore gene-longevity associations: the ECHA study.** *Eur J Hum Genet* 2008, **16(2)**:236-242.

31. Skytthe A, Valensin S, Jeune B, Cevenini E, Balard F, Beekman M, Bezrukov V, Blanche H, Bolund L, Broczek K, Carru C, Christensen K, Christiansen L, Collerton JC, Cotichini R, de Craen AJ, Dato S, Davies K, De Benedictis G, Deiana L, Flachsbart F, Gampe J, Gilbault C, Gonos ES, Haimes E, Hervonen A, Hurme MA, Janiszewska D, Jylhä M, Kirkwood TB, Kristensen P, Laiho P, Leon A, Marchisio A, Masciulli R, Nebel A, Passarino G, Pelicci G, Peltonen L, Perola M, Poulain M, Rea IM, Remacle J, Robine JM, Schreiber S, Scurti M, Sevini F, Sikora E, Skouteri A, Slagboom PE, Spazzafumo L, Stazi MA, Toccaceli V, Toussaint O, Törnwall O, Vaupel JW, Voutetakis K, Franceschi C, GEHA consortium: **Design, recruitment, logistics, and data management of the GEHA (Genetics of Healthy Ageing) project.** *Exp Gerontol* 2011, **46(11)**:934-945.

32. Wang S, Zhao H: **Sample size needed to detect gene-gene interactions using association designs.** *Am J Epidemiol* 2003, **158(9)**:899-914.

33. Christensen K, Johnson TE, Vaupel JW: **The quest for genetic determinants of human longevity: Challenges and insights.** *Nat Rev Genet* 2006, **7(6)**:436-448.

34. Yashin AI, De Benedictis G, Vaupel JW, Tan Q, Andreev KF, Iachine IA, Bonafe M, DeLuca M, Valensin S, Carotenuto L, Franceschi C: **Genes, demography, and life span: the contribution of demographic data in genetic studies on aging and longevity.** *Am J Hum Genet* 1999, **65(4)**:1178-1193.

35. Cardelli M, Cavallone L, Marchegiani F, Oliveri F, Dato S, Montesanto A, Lescai F, Lisa R, De Benedictis G, Franceschi C: **A genetic-demographic approach reveals male-specific association between survival and tumor necrosis factor (A/G)-308 polymorphism.** *J Gerontol A Biol Sci Med Sci* 2008, **63(5)**:454-460.

36. Hoh J, Ott J: **Mathematical multi-locus approaches to localizing complex human trait genes.** *Nat Rev Genet* 2003, **4(9)**:701-709.

37. Clayton DG, Chapman JM, Cooper JD: **Use of unphased multilocus genotype data in indirect association studies.** *Genet Epidemiol* 2004, **27**:415-428.

38. Liu Y, Xu H, Chen S, Chen X, Zhang Z, Zhu Z, Qin X, Hu L, Zhu J, Zhao GP, Kong X: **Genome-wide interaction-based association analysis identified multiple new susceptibility Loci for common diseases.** *PLoS Genet* 2011, **7(3)**:e1001338.

39. Ritchie MD, Hahn LW, Roodi N, Bailey LR, Dupont WD, Parl FF, Moore JH: **Multifactor-dimensionality reduction reveals high-order interactions among estrogen-metabolism genes in sporadic breast cancer.** *Am J Hum Genet* 2001, **69(1)**:138-147.

40. Akey J, Jin L, Xiong M: **Haplotypes vs single marker linkage disequilibrium tests: what do we gain?** *Eur J Hum Genet* 2001, **9(4)**:291-300.

41. Clark AG: **The role of haplotypes in candidate gene studies.** *Genet Epidemiol* 2004, **27(4)**:321-333.

42. Schaid DJ: **Linkage disequilibrium testing when linkage phase is unknown.** *Genetics* 2004, **166(1)**:505-512.

43. Schaid DJ, Rowland CM, Tines DE, Jacobson RM, Poland GA: **Score tests for association between traits and haplotypes when linkage phase is ambiguous.** *Am J Hum Genet* 2002, **70(2)**:425-434.

44. Tan Q, Christiansen L, Bathum L, Zhao JH, Yashin AI, Vaupel JW, Christensen K, Kruse TA: **Estimating haplotype relative risks on human survival in population-based association studies.** *Hum Hered* 2005, **59(2)**:88-97.

45. Tan Q, De Benedictis G, Yashin AI, Bathum L, Christiansen L, Dahlgaard J, Frizner N, Vach W, Vaupel JW, Christensen K, Kruse TA: **Assessing genetic association with human survival at multi-allelic loci.** *Biogerontology* 2004, **5(2)**:89-97.

46. Lunetta KL, D'Agostino RB Sr, Karasik D, Benjamin EJ, Guo CY, Govindaraju R, Kiel DP, Kelly-Hayes M, Massaro JM, Pencina MJ, Seshadri S, Murabito JM: **Genetic correlates of longevity and selected age-related phenotypes: a genome-wide association study in the Framingham Study.** *BMC Med Genet* 2007, **8(Suppl 1)**:S13.

47. Newman AB, Walter S, Lunetta KL, Garcia ME, Slagboom PE, Christensen K, Arnold AM, Aspelund T, Aulchenko YS, Benjamin EJ, Christiansen L, D'Agostino RB Sr, Fitzpatrick AL, Franceschini N, Glazer NL, Gudnason V, Hofman A, Kaplan R, Karasik D, Kelly-Hayes M, Kiel DP, Launer LJ, Marciante KD, Massaro JM, Miljkovic I, Nalls MA, Hernandez D, Psaty BM, Rivadeneira F, Rotter J, Seshadri S, Smith AV, Taylor KD, Tiemeier H, Uh HW, Uitterlinden AG, Vaupel JW, Westendorp RG, Walston J, Harris TB, Lumley T, van Duijn CM, Murabito JM: **A meta-analysis of four genome-wide association studies of survival to age 90 years or older: the Cohorts for Heart and Aging Research in Genomic Epidemiology Consortium.** *J Gerontol A Biol Sci Med Sci* 2010, **65(5)**:478-487.

48. Walter S, Atzmon G, Demerath EW, Garcia ME, Kaplan RC, Kumari M, Lunetta KL, Milaneschi Y, Tanaka T, Tranah GJ, Völker U, Yu L, Arnold A, Benjamin EJ, Biffar R, Buchman AS, Boerwinkle E, Couper D, De Jager PL, Evans DA, Harris TB, Hoffmann W, Hofman A, Karasik D, Kiel DP, Kocher T, Kuningas M, Launer LJ, Lohman KK, Lutsey PL, Mackenbach J, Marciante K, Psaty BM, Reiman EM, Rotter JI, Seshadri S, Shardell MD, Smith AV, van Duijn C, Walston J, Zillikens MC, Bandinelli S, Baumeister SE, Bennett DA, Ferrucci L, Gudnason V, Kivimaki M, Liu Y, Murabito JM, Newman AB, Tiemeier H, Franceschini N: **A genome-wide association study of aging.** *Neurobiol Aging* 2011, **32(11)**:2109, e15-28.

49. Deelen J, Uh HW, Monajemi R, van Heemst D, Thijssen PE, Böhringer S, van den Akker EB, de Craen AJ, Rivadeneira F, Uitterlinden AG, Westendorp RG, Goeman JJ, Slagboom PE, Houwing-Duistermaat JJ, Beekman M: **Gene set analysis of GWAS data for human longevity highlights the relevance of the insulin/IGF-1 signaling and telomere maintenance pathways.** *Age* 2011, doi:10.1007/s11357-011-9340-3.

50. Nebel A, Kleindorp R, Caliebe A, Nothnagel M, Blanché H, Junge O, Wittig M, Ellinghaus D, Flachsbart F, Wichmann HE, Meitinger T, Nikolaus S, Franke A, Krawczak M, Lathrop M, Schreiber S: **A genome-wide association study confirms APOE as the major gene influencing survival in long-lived individuals.** *Mech Ageing Dev* 2011, **132(6-7)**:324-330.

51. Malovini A, Illario M, Iaccarino G, Villa F, Ferrario A, Roncarati R, Anselmi CV, Novelli V, Cipolletta E, Leggiero E, Orro A, Rusciano MR, Milanesi L, Maione AS, Condorelli G, Bellazzi R, Puca AA: **Association study on long-living individuals from Southern Italy identifies rs10491334 in the CAMKIV gene that regulates survival proteins.** *Rejuvenation Res* 2011, **14(3)**:283-291.

52. Sebastiani P, Solovieff N, Dewan AT, Walsh KM, Puca A, Hartley SW, Melista E, Andersen S, Dworkis DA, Wilk JB, Myers RH, Steinberg MH, Montano M, Baldwin CT, Hoh J, Perls TT: **Genetic signatures of exceptional longevity in humans.** *PLoS One* 2012, **7(1)**:e29848.

53. Nebel A, Flachsbart F, Till A, Caliebe A, Blanché H, Arlt A, Häsler R, Jacobs G, Kleindorp R, Franke A, Shen B, Nikolaus S, Krawczak M, Rosenstiel P, Schreiber S: **A functional EXO1 promoter variant is associated with prolonged life expectancy in centenarians.** *Mech Ageing Dev* 2009, **130(10)**:691-699.

54. Pawlikowska L, Hu D, Huntsman S, Sung A, Chu C, Chen J, Joyner AH, Schork NJ, Hsueh WC, Reiner AP, Psaty BM, Atzmon G, Barzilai N, Cummings SR, Browner WS, Kwok PY, Ziv E: **Association of common genetic variation in the insulin/IGF1 signaling pathway with human longevity.** *Aging Cell* 2009, **8(4)**:460-472.

55. Flachsbart F, Franke A, Kleindorp R, Caliebe A, Blanché H, Schreiber S, Nebel A: **Investigation of genetic susceptibility factors for human longevity - a targeted nonsynonymous SNP study.** *Mutat Res* 2010, **694(1-2)**:13-19.

56. Soerensen M, Dato S, Tan Q, Thinggaard M, Kleindorp R, Beekman M, Suchiman HE, Jacobsen R, McGue M, Stevnsner T, Bohr VA, de Craen AJ, Westendorp RG, Schreiber S, Slagboom PE, Nebel A, Vaupel JW, Christensen K, Christiansen L: **Evidence from case-control and longitudinal studies supports associations of genetic variation in APOE, CETP, and IL6 with human longevity.** *Age* 2012, doi:10.1007/s11357-011-9373-7.

57. Carlson CS, Eberle MA, Rieder MJ, Yi Q, Kruglyak L, Nickerson DA: **Selecting a maximally informative set of single-nucleotide polymorphisms for association analyses using linkage disequilibrium.** *Am J Hum Genet* 2004, **74**(1):106-120.

58. Donmez G, Guarente L: **Aging and disease: connections to sirtuins.** *Aging Cell* 2010, **9**(2):285-290.

59. Partridge L: **Some highlights of research on aging with invertebrates, 2010.** *Aging Cell* 2011, **10**(1):5-9.

60. Trinei M, Berniakovich I, Beltrami E, Migliaccio E, Fassina A, Pelicci P, Giorgio M: **P66Shc signals to age.** *Aging (Albany NY)* 2009, **1**(6):503-510.

61. Johnson TE: **Caenorhabditis elegans 2007: the premier model for the study of aging.** *Exp Gerontol* 2008, **43**(1):1-4.

62. Kenyon CJ: **The genetics of ageing.** *Nature* 2010, **464**(7288):504-512.

63. Arking DE, Krebsova A, Macek M Sr, Macek M Jr, Arking A, Mian IS, Fried L, Hamosh A, Dey S, McIntosh I, Dietz HC: **Association of human aging with a functional variant of klotho.** *Proc Natl Acad Sci USA* 2002, **99**(2):856-861.

64. Soerensen M, Dato S, Christensen K, McGue M, Stevnsner T, Bohr VA, Christiansen L: **Replication of an association of variation in the FOXO3A gene with human longevity using both case-control and longitudinal data.** *Aging Cell* 2010, **9**(6):1010-1017.

65. Rose G, Crocco P, De Rango F, Montesanto A, Passarino G: **Further support to the uncoupling-to-survive theory: the genetic variation of human UCP genes is associated with longevity.** *PLoS One* 2011, **6**(12):29650.

66. Rose G, Crocco P, D'Aquila P, Montesanto A, Bellizzi D, Passarino G: **Two variants located in the upstream enhancer region of human UCP1 gene affect gene expression and are correlated with human longevity.** *Exp Gerontol* 2011, **46**(11):897-904.

67. Bonafè M, Barbi C, Storci G, Salvioli S, Capri M, Olivieri F, Valensin S, Monti D, Gonos ES, De Benedictis G, Franceschi C: **What studies on human longevity tell us about the risk for cancer in the oldest old: data and hypotheses on the genetics and immunology of centenarians.** *Exp Gerontol* 2002, **37**(10-11):1263-1271.

68. Slagboom PE, Droog S, Boomsma DI: **Genetic determination of telomere size in humans: a twin study of three age groups.** *Am J Hum Genet* 1994, **55**(5):876-882.

69. Beekman M, Nederstigt C, Suchiman HE, Kremer D, van der Breggen R, Lakenberg N, Alemayehu WG, de Craen AJ, Westendorp RG, Boomsma DI, de Geus EJ, Houwing-Duistermaat JJ, Heijmans BT, Slagboom PE: **Genome-wide association study (GWAS)-identified disease risk alleles do not compromise human longevity.** *Proc Natl Acad Sci USA* 2010, **107**(42):18046-18049.

70. Bergman A, Atzmon G, Ye K, MacCarthy T, Barzilai N: **Buffering mechanisms in aging: a systems approach toward uncovering the genetic component of aging.** *PLoS Comput Biol* 2007, **3**(8):e170.

71. Fontana L, Partridge L, Longo VD: **Extending healthy life span-from yeast to humans.** *Science* 2010, **328**(5976):321-326.

72. Tatar M, Bartke A, Antebi A: **The endocrine regulation of aging by insulin-like signals.** *Science* 2003, **299**(5611):1346-1351.

73. Murphy CT, McCarroll SA, Bargmann CI, Fraser A, Kamath RS, Ahringer J, Li H, Kenyon C: **Genes that act downstream of DAF-16 to influence the lifespan of Caenorhabditis elegans.** *Nature* 2003, **424**(6946):277-283.

74. Bjedov I, Partridge L: **A longer and healthier life with TOR down-regulation: genetics and drugs.** *Biochem Soc Trans* 2011, **39**(2):460-465.

75. Gerdes LU, Jeune B, Ranberg KA, Nybo H, Vaupel JW: **Estimation of apolipoprotein E genotype-specific relative mortality risks from the distribution of genotypes in centenarians and middle-aged men: apolipoprotein. E gene is a "frailty gene," not a "longevity gene".** *Genet Epidemiol* 2000, **19**(3):202-210.

76. Rose G, Dato S, Altomare K, Bellizzi D, Garasto S, Greco V, Passarino G, Feraco E, Mari V, Barbi C, BonaFe M, Franceschi C, Tan Q, Boiko S, Yashin AI, De Benedictis G: **Variability of the SIRT3 gene, human silent information regulator Sir2 homologue, and survivorship in the elderly.** *Exp Gerontol* 2003, **38**(10):1065-1070.

77. Soerensen M, Dato S, Christensen K, McGue M, Stevnsner T, Bohr VA, Christiansen L: **Replication of an association of variation in the FOXO3A gene with human longevity using both case-control and longitudinal data.** *Aging Cell* 2010, **9**:1010-1017.

78. Bonafè M, Olivieri F: **Genetic polymorphism in long-lived people: cues for the presence of an insulin/IGF-pathway-dependent network affecting human longevity.** *Mol Cell Endocrinol* 2009, **299**(1):118-123.

79. Chung WH, Dao RL, Chen LK, Hung SI: **The role of genetic variants in human longevity.** *Ageing Res Rev* 2010, **9**(Suppl 1):S67-S78.

80. Invidia L, Salvioli S, Altilia S, Pierini M, Panourgia MP, Monti D, De Rango F, Passarino G, Franceschi C: **The frequency of Klotho KL-VS polymorphism in a large Italian population, from young subjects to centenarians, suggests the presence of specific time windows for its effect.** *Biogerontology* 2010, **11**(1):67-73.

81. Slagboom PE, Beekman M, Passtoors WM, Deelen J, Vaarhorst AA, Boer JM, van den Akker EB, van Heemst D, de Craen AJ, Maier AB, Rozing M, Mooijaart SP, Heijmans BT, Westendorp RG: **Genomics of human longevity.** *Philos Trans R Soc Lond B Biol Sci* 2011, **366**(1561):35-42.

82. Bathum L, Christiansen L, Tan Q, Vaupel J, Jeune B, Christensen K: **No evidence for an association between extreme longevity and microsomal transfer protein polymorphisms in a longitudinal study of 1651 nonagenarians.** *Eur J Hum Genet* 2005, **13**(10):1154-1158.

83. Chanock SJ, Manolio T, Boehnke M, Boerwinkle E, Hunter DJ, Thomas G, Hirschhorn JN, Abecasis G, Altshuler D, Bailey-Wilson JE, Brooks LD, Cardon LR, Daly M, Donnelly P, Fraumeni JF Jr, Freimer NB, Gerhard DS, Gunter C, Guttmacher AE, Guyer MS, Harris EL, Hoh J, Hoover R, Kong CA, Merikangas KR, Morton CC, Palmer LJ, Phimister EG, Rice JP, Roberts J, Rotimi C, Tucker MA, Vogan KJ, Wacholder S, Wijsman EM, Winn DM, Collins FS: **Replicating genotype-phenotype associations.** *Nature* 2007, **447**(7145):655-660.

84. Kerber RA, O'Brien E, Cawthon RM: **Gene expression profiles associated with aging and mortality in humans.** *Aging Cell* 2009, **8**:239-250.

85. Foger N, Rangell L, Danilenko DM, Chan AC: **Requirement for coronin 1 in T lymphocyte trafficking and cellular homeostasis.** *Science* 2006, **313**(5788):839-842.

86. Mathers JC: **Nutritional modulation of ageing: genomic and epigenetic approaches.** *Mech Ageing Dev* 2006, **127**:584-589.

87. Frisard M, Ravussin E: **Energy metabolism and oxidative stress: impact on the metabolic syndrome and the aging process.** *Endocrine* 2006, **29**(1):27-32.

88. Frisard MI, Broussard A, Davies SS, Roberts LJ, Rood J, de Jonge L, Fang X, Jazwinski SM, Deutsch WA, Ravussin E: **Aging, resting metabolic rate, and oxidative damage: results from the Louisiana.** *J Gerontol A Biol Sci Med Sci* 2007, **62**(7):752-759.

89. Otera H, Mihara K: **Molecular mechanisms and physiologic functions of mitochondrial dynamics.** *J Biochem* 2011, **149**(3):241-251.

90. Scheffler IE: **Mitochondria make a come back.** *Adv Drug Deliv Rev* 2001, **49**(1-2):3-26.

91. Santos SD, Manadas B, Duarte CB, Carvalho AL: **Proteomic analysis of an interactome for long-form AMPA receptor subunits.** *J Proteome Res* 2010, **9**(4):1670-1682.

92. Salvioli S, Capri M, Santoro A, Raule N, Sevini F, Lukas S, Lanzarini C, Monti D, Passarino G, Rose G, De Benedictis G, Franceschi C: **The impact of mitochondrial DNA on human lifespan: a view from studies on centenarians.** *Biotechnol J* 2008, **3**(6):740-749.

93. Chomyn A, Attardi G: **MtDNA mutations in aging and apoptosis.** *Biochem Biophys Res Commun* 2003, **304**(3):519-529.

94. Kivisild T, Bamshad MJ, Kaldma K, Metspalu M, Metspalu E, Reidla M, Laos S, Parik J, Watkins WS, Dixon ME, Papiha SS, Mastana SS, Mir MR, Ferak V, Villems R: **Deep common ancestry of indian and western-Eurasian mitochondrial DNA lineages.** *Curr Biol* 1999, **9**(22):1331-1334.

95. Mishmar D, Ruiz-Pesini E, Golik P, Macaulay V, Clark AG, Hosseini S, Brandon M, Easley K, Chen E, Brown MD, Sukernik RI, Olckers A, Wallace DC: **Natural selection shaped regional mtDNA variation in humans.** *Proc Natl Acad Sci USA* 2003, **100**(1):171-176.

96. Torroni A, Petrozzi M, D'Urbano L, Sellitto D, Zeviani M, Carrara F, Carducci C, Leuzzi V, Carelli V, Barboni P, De Negri A, Scozzari R: **Haplotype and phylogenetic analyses suggest that one European- specific mtDNA background plays a role in the expression of Leber hereditary optic neuropathy by increasing the penetrance of the primary mutations 11778 and 14484.** *Am J Hum Genet* 1997, **60**(5):1107-1121.

97. Wallace DC: **A mitochondrial paradigm of metabolic and degenerative diseases, aging, and cancer: A dawn for evolutionary medicine.** *Annu Rev Genet* 2005, **39**:359-407.

98. De Benedictis G, Rose G, Carrieri G, De Luca M, Falcone E, Passarino G, Bonafe M, Monti D, Baggio G, Bertolini S, Mari D, Mattace R, Franceschi C: **Mitochondrial DNA inherited variants are associated with successful aging and longevity in humans.** *FASEB J* 1999, **13**(12):1532-1536.

99. Zhang J, Asin-Cayuela J, Fish J, Michikawa Y, Bonafe M, Olivieri F, Passarino G, De Benedictis G, Franceschi C, Attardi G: **Strikingly higher frequency in centenarians and twins of mtDNA mutation causing remodeling of replication origin in leukocytes.** *Proc Natl Acad Sci USA* 2003, **100**(3):1116-1121.

100. Rose G, Passarino G, Scornaienchi V, Romeo G, Dato S, Bellizzi D, Mari V, Feraco E, Maletta R, Bruni A, Franceschi C, De Benedictis G: **The mitochondrial DNA control region shows genetically correlated levels of heteroplasmy in leukocytes of centenarians and their offspring.** *BMC Genomics* 2007, **8**:293.

101. Rose G, Romeo G, Dato S, Crocco P, Bruni AC, Hervonen A, Majamaa K, Sevini F, Franceschi C, Passarino G: **Somatic point mutations in mtDNA control region are influenced by genetic background and associated with healthy aging: a GEHA study.** *PLoS One* 2010, **5**(10):13395.

102. Ryan MT, Hoogenraad NJ: **Mitochondrial-nuclear communications.** *Annu Rev Biochem* 2007, **76**:701-722.

103. Jazwinski SM: **The retrograde response: When mitochondrial quality control is not enough.** *Biochim Biophys Acta* 2012, doi:10.1016/j. bbamcr.2012.02.010.

104. Bellizzi D, Cavalcante P, Taverna D, Rose G, Passarino G, Salvioli S, Franceschi C, De Benedictis G: **Gene expression of cytokines and cytokine receptors is modulated by the common variability of the mitochondrial DNA in cybrid cell lines.** *Genes Cells* 2006, **11**(8):883-891.

105. Bellizzi D, Taverna D, D'Aquila P, De Blasi S, De Benedictis G: **Mitochondrial DNA variability modulates mRNA and intra-mitochondrial protein levels of HSP60 and HSP75: experimental evidence from cybrid lines.** *Cell Stress Chaperones* 2009, **14**(3):265-271.

106. Campisi J, d'Adda di Fagagna F: **Cellular senescence: when bad things happen to good cells.** *Nat Rev Mol Cell Biol* 2007, **8**(9):729-740.

107. Oeseburg H, de Boer RA, van Gilst WH, van der Harst P: **Telomere biology in healthy aging and disease.** *Pflugers Arch* 2010, **459**(2):259-268.

108. Blasco MA: **Telomeres and human disease: ageing, cancer and beyond.** *Nat Rev Genet* 2005, **6**(8):611-622.

109. Shay JW, Bacchetti S: **A survey of telomerase activity in human cancer.** *Eur J Cancer* 1997, **33**(5):787-791.

110. Cassar L, Li H, Pinto AR, Nicholls C, Bayne S, Liu JP: **Bone morphogenetic protein-7 inhibits telomerase activity, telomere maintenance, and cervical tumor growth.** *Cancer Res* 2008, **68**(22):9157-9166.

111. Terry DF, Nolan VG, Andersen SL, Perls TT, Cawthon R: **Association of longer telomeres with better health in centenarians.** *J Gerontol A Biol Sci Med Sci* 2008, **63**(8):809-812.

112. Atzmon G, Cho M, Cawthon RM, Budagov T, Katz M, Yang X, Siegel G, Bergman A, Huffman DM, Schechter CB, Wright WE, Shay JW, Barzilai N, Govindaraju DR, Suh Y: **Evolution in health and medicine Sackler colloquium: Genetic variation in human telomerase is associated with telomere length in Ashkenazi centenarians.** *Proc Natl Acad Sci USA* 2010, , Suppl 1: 1710-1717.

113. Ravaglia G, Morini P, Forti P, Maioli F, Boschi F, Bernardi M, Gasbarrini G: **Anthropometric characteristics of healthy Italian nonagenarians and centenarians.** *Br J Nutr* 1997, **77**(1):9-17.

114. Poon LW, Jazwinski M, Green RC, Woodard JL, Martin P, Rodgers WL, Johnson MA, Hausman D, Arnold J, Davey A, Batzer MA, Markesbery WR, Gearing M, Siegler IC, Reynolds S, Dai J: **Methodological considerations in studying centenarians: lessons learned from the Georgia centenarian studies.** *Annual Review of Gerontology and Geriatrics* 2007, **27**(1):231-264.

115. Samuelsson SM, Alfredson BB, Hagberg B, Samuelsson G, Nordbeck B, Brun A, Gustafson L, Risberg J: **The Swedish Centenarian Study: a multidisciplinary study of five consecutive cohorts at the age of 100.** *Int J Aging Hum Dev* 1997, **45**(3):223-253.

116. Martin P, Poon LW, Clayton GM, Lee HS, Fulks JS, Johnson MA: **Personality, life events, and coping in the oldest-old.** *Int J Aging Hum Dev* 1992, **34**(1):19-30.

117. Capurso A, Resta F, D'Amelio A, Gaddi A, Daddato S, Galletti C, Trabucchi M, Boffelli S, Rozzini R, Motta L, Rapisarda R, Receputo G, Mattace R, Motta M, Pansini L, Masotti G, Marchionni N, Petruzzi E, Bertolini S, Agretti M, Costelli P, Mari D, Duca F, Ferrazzi P, Bosi E, Manzoni M, Tomasello FB, Salvioli G, Baldelli MV, Neri M, Franceschi C, Cossarizza A, Monti D, Varricchio M, Gambardella A, Paolisso G, Baggio G, DallaVestra M, Donazzan S, Sangiorgi GB, Barbagallo M, Frada G, Passeri M, Fagnoni F, Sansoni P, Senin U, Cherubini A, Polidori MC, Marigliano V, Bauco C, Cacciafesta M, Forconi S, Guerrini M, Boschi S, Fabris F, Cappa G, Ferrario E,

Giarelli L, Cavalieri F, Stanta G: **Epidemiological and socioeconomic aspects of Italian centenarians.** *Arch Gerontol Geriatr* 1997, **25**:149-157.

118. Poon LW: **The Georgia Centenarian Study.** *Int J Aging Hum Dev* 1992, **34**(1):1-17.

119. Franceschi C, Motta L, Valensin S, Rapisarda R, Franzone A, Berardelli M, Motta M, Monti D, Bonafe' M, Ferrucci L, Deiana L, Pes GM, Carru C, Desole MS, Barbi C, Sartoni G, Gemelli C, Lescai F, Olivieri F, Marchegiani F, Cardelli M, Cavallone L, Gueresi P, Cossarizza A, Troiano L, Pini G, Sansoni P, Passeri G, Lisa R, Spazzafumo L, Amadio L, Giunta S, Stecconi R, Morresi R, Viticchi C, Mattace R, De Benedictis G, Baggio G: **Do men and women follow different trajectories to reach extreme longevity? Italian Multicenter Study on Centenarians (IMUSCE).** *Aging* 2000, **12**(2):77-84.

120. Jeune B: **Living longer-but better?** *Aging Clin Exp Res* 2002, **14**(2):72-93.

121. Fried LP, Tangen CM, Walston J, Newman AB, Hirsch C, Gottdiener J, Seeman T, Tracy R, Kop WJ, Burke G, McBurnie MA: **Frailty in older adults: evidence for a phenotype.** *J Gerontol A Biol Sci Med Sci* 2001, **56**(3):146-156.

122. Fried LP, Ferrucci L, Darer J, Williamson JD, Anderson G: **Untangling the concepts of disability, frailty, and comorbidity: implications for improved targeting and care.** *J Gerontol A Biol Sci Med Sci* 2004, **59**(3):255-263.

123. Bortz WM: **A conceptual framework of frailty: a review.** *J Gerontol A Biol Sci Med Sci* 2002, **57**(5):M283-M288.

124. Mitnitski AB, Graham JE, Mogilner AJ, Rockwood K: **Frailty, fitness and late-life mortality in relation to chronological and biological age.** *BMC Geriatr* 2002, **2**:1.

125. Jones DM, Song X, Rockwood K: **Operationalizing a frailty index from a standardized comprehensive geriatric assessment.** *J Am Geriatr Soc* 2004, **52**(11):1929-1933.

126. Jones D, Song X, Mitnitski A, Rockwood K: **Evaluation of a frailty index based on a comprehensive geriatric assessment in a population based study of elderly Canadians.** *Aging Clin Exp Res* 2005, **17**(6):465-471.

127. Walston J, Hadley EC, Ferrucci L, Guralnik JM, Newman AB, Studenski SA, Ershler WB, Harris T, Fried LP: **Research agenda for frailty in older adults: toward a better understanding of physiology and etiology: summary from the American Geriatrics Society/National Institute on Aging Research Conference on Frailty in Older Adults.** *J Am Geriatr Soc* 2006, **54**(6):991-1001.

128. Jeune B, Skytthe A, Cournil A, Greco V, Gampe J, Berardelli M, Andersen-Ranberg K, Passarino G, Debenedictis G, Robine JM: **Handgrip strength among nonagenarians and centenarians in three European regions.** *J Gerontol A Biol Sci Med Sci* 2006, **61**(7):707-712.

129. Montesanto A, Lagani V, Martino C, Dato S, De Rango F, Berardelli M, Corsonello A, Mazzei B, Mari V, Lattanzio F, Conforti D, Passarino G: **A novel, population-specific approach to define frailty.** *Age* 2010, **32**(3):385-395.

130. Dato S, Montesanto A, Lagani V, Jeune B, Christensen K, Passarino G: **Frailty phenotypes in the elderly based on cluster analysis: a longitudinal study of two Danish cohorts. Evidence for a genetic influence on frailty.** *Age* 2011, doi:10.1007/s11357-011-9257-x.

131. Wolffe AP, Matzke MA: **Epigenetics: regulation through repression.** *Science* 1999, **286**(5439):481-486.

132. Fraga MF: **Genetic and epigenetic regulation of aging.** *Curr Opin Immunol* 2009, **21**(4):446-453.

133. Fraga MF, Esteller M: **Towards the human cancer epigenome: a first draft of histone modifications.** *Cell Cycle* 2005, **4**(10):1377-1381.

134. Bjornsson HT, Sigurdsson MI, Fallin MD, Irizarry RA, Aspelund T, Cui H, Yu W, Rongione MA, Ekström TJ, Harris TB, Launer LJ, Eiriksdottir G, Leppert MF, Sapienza C, Gudnason V, Feinberg AP: **Intra-individual change over time in DNA methylation with familial clustering.** *JAMA* 2008, **299**(24):2877-2883.

135. Bollati V, Schwartz J, Wright R, Litonjua A, Tarantini L, Suh H, Sparrow D, Vokonas P, Baccarelli A: **Decline in genomic DNA methylation through aging in a cohort of elderly subjects.** *Mech Ageing Dev* 2009, **130**(4):234-239.

136. Wilson VL, Smith RA, Ma S, Cutler RG: **Genomic 5-methyldeoxycytidine decreases with age.** *J Biol Chem* 1987, **262**(21):9948-9951.

137. Oakes CC, Smiraglia DJ, Plass C, Trasler JM, Robaire B: **Aging results in hypermethylation of ribosomal DNA in sperm and liver of male rats.** *Proc Natl Acad Sci USA* 2003, **100**(4):1775-1780.

138. Richardson B: **Impact of aging on DNA methylation.** *Ageing Res Rev* 2003, **2**(3):245-261.

139. Fuke C, Shimabukuro M, Petronis A, Sugimoto J, Oda T, Miura K, Miyazaki T, Ogura C, Okazaki Y, Jinno Y: **Age related changes in 5-methylcytosine content in human peripheral leukocytes and placentas: an HPLC-based study.** *Ann Hum Genet* 2004, **68**(Pt 3):196-204.

140. Fraga MF, Esteller M: **Epigenetics and aging: the targets and the marks.** *Trends Genet* 2007, **23**(8):413-418.

141. Ling C, Del Guerra S, Lupi R, Rönn T, Granhall C, Luthman H, Masiello P, Marchetti P, Groop L, Del Prato S: **Epigenetic regulation of PPARGC1A in human type 2 diabetic islets and effect on insulin secretion.** *Diabetologia* 2008, **51**(4):615-622.

142. Arai T, Kasahara I, Sawabe M, Honma N, Aida J, Tabubo K: **Role of methylation of the hMLH1 gene promoter in the development of gastric and colorectal carcinoma in the elderly.** *Geriatr Gerontol Int* 2010, **10**(Suppl 1):S207-S212.

143. Lee J, Jeong DJ, Kim J, Lee S, Park JH, Chang B, Jung SI, Yi L, Han Y, Yang Y, Kim KI, Lim JS, Yang I, Jeon S, Bae DH, Kim CJ, Lee MS: **The anti-aging gene KLOTHO is a novel target for epigenetic silencing in human cervical carcinoma.** *Mol Cancer* 2010, **9**:109.

144. Bellizzi D, D'Aquila P, Montesanto A, Corsonello A, Mari V, Mazzei B, Lattanzio F, Passarino G: **Global DNA methylation in old subjects is correlated with frailty.** *Age (Dordr)* 2012, **34**(1):169-179.

doi:10.1186/1742-4933-9-6
Cite this article as: Montesanto *et al.*: **Epidemiological, genetic and epigenetic aspects of the research on healthy ageing and longevity.** *Immunity & Ageing* 2012 **9**:6.

Ferrario et al. Immunity & Ageing 2012, 9:7
http://www.immunityageing.com/content/9/1/7

IMMUNITY & AGEING

REVIEW

The application of genetics approaches to the study of exceptional longevity in humans: potential and limitations

Anna Ferrario[1], Francesco Villa[1], Alberto Malovini[2], Fiorella Araniti[1] and Annibale A Puca[1,3*]

Abstract

The average life-span of the population of industrialized countries has improved enormously over the last decades. Despite evidence pointing to the role of food intake in modulating life-span, exceptional longevity is still considered primarily an inheritable trait, as pointed out by the description of families with centenarian clusters and by the elevated relative probability of siblings of centenarians to become centenarians themselves. However, rather than being two separate concepts, the genetic origin of exceptional longevity and the more recently observed environment-driven increase in the average age of the population could possibly be explained by the same genetic variants and environmentally modulated mechanisms (caloric restriction, specific nutrients). In support of this hypothesis, polymorphisms selected for in the centenarian population as a consequence of demographic pressure have been found to modulate cellular signals controlled also by caloric restriction. Here, we give an overview of the recent findings in the field of the genetics of human exceptional longevity, of how some of the identified polymorphisms modulate signals also influenced by food intake and caloric restriction, of what in our view have been the limitations of the approaches used over the past years to study genetics (sib-pair-, candidate gene association-, and genome-wide association-studies), and briefly of the limitations and the potential of the new, high-throughput, next-generation sequencing techniques applied to exceptional longevity.

Keywords: Aging, Centenarians, Longevity

Mechanisms of longevity

Life expectancy in the US 1900 Birth Cohort Study was found to be 51.5 years for males and 58.3 years for females, and currently 1/10,000 individuals reach 100 years of age: this prevalence is quickly changing and will probably soon approach 1/5,000 [1]. The increased ability to reach 100 years old in industrialized countries over the last 160 years most likely reflects a rise in life expectancy — quantified as 3 months/year for females — as a consequence of improvements in diet and a reduced exposure to infection and inflammation [2]. In favour of diet as a modulator of longevity, the Elderly Prospective Cohort Study (EPIC) identified a reduced overall mortality among the elderly consuming a modified Mediterranean

diet in which saturated fatty acids were substituted for monounsaturated ones [3].

Centenarians, despite being exposed to the same environmental conditions as members of the average population, manage to live much longer; moreover, as a consequence of demographic selection, centenarians have a compression of morbidity and mortality towards the end of their life-span [4]. Genetically, this compression in morbidity and mortality is correlated with the enrichment of protective alleles and the depletion of detrimental ones. These alleles run in families, as shown by the familiar clustering of exceptional longevity. It has been estimated that genetic variants account for at least 25% of human life-span, and for even a larger proportion in individuals living to extreme age [5,6].

The potential overlap of hits for environmentally and genetically mediated predisposition for extreme longevity in centenarians is highlighted by the association of genetic variants of genes that regulate, or that are

* Correspondence: puca@longevita.org
[1]IRCCS Multimedica, Via Fantoli 16/15, 20138 Milan, Italy
[3]Università degli Studi di Salerno, Via S. Allende, 84081 Baronissi, Italy
Full list of author information is available at the end of the article

regulated by, nutrient metabolism, such as apolipoprotein E (APOE) and Forkhead box O3A (FOXO3A) [7]. Inheritance of the longevity phenotype is underlined by the low all-cause and cardiovascular-disease mortality rates observed in offspring of centenarians when compared with an age-matched population [8]. The study of centenarian offspring has revealed biomarkers of longevity, like low serum levels of heat shock proteins (HSP), large lipid particle sizes, and high membrane palmitoleic acid paired with a low peroxidation index [9-11]. In addition, centenarians have high glucose tolerance and insulin action, and low heart-rate variability (HRV), which contrasts with the decline observed in control populations [12,13].

Genetic variants that modulate human longevity should also modulate cellular pathways that control key aspects of the aging process, such as oxidative stress-induced apoptosis (RAS/ERK pathway), DNA repair (NF-KB1 and hTERT), senescence (p53), mitochondrial biogenesis (AMPK) and cell survival (PI3K/AKT pathway). Many of these pathways cross-talk with each other and are finely regulated in order to obtain the best trade-off between advantages and disadvantages of inducing either cell survival or apoptosis/senescence. The best trade-off is mostly tissue specific and can change with the degree of the tissue's differentiation (stem versus differentiated cells) and health status (healthy, proliferative, degenerative or ischemic). It is plausible that genetic variants that impact on human longevity affect genes that are expressed selectively in some tissues and/or in specific stages of differentiation. Although aging has been considered an evolutionary adaptation for fighting cancer through the activation of processes like senescence, there are signals that are able to activate cell-specific responses, e.g. induction of apoptosis by adenosine monophosphate-activated protein kinase (AMPK) in cancer cells, while inducing survival in healthy cells. AMPK recognizes high levels of AMP, and is activated by caloric restriction, physical exercise, metformin, essential amino acids and alpha-lipoic acid [14]. Furthermore, AMPK induces mitochondrial biogenesis, autophagy and beta-oxidation of free fatty acids [15-17]. Reduction of fatty acid beta-oxidation promotes diabetes, obesity and, ultimately, aging.

How can these cellular signals be altered without producing side effects? The answer could be found in the centenarian genome through the identification of genetic variants that have been selected or dropped for their role in human health. AMPK signalling is an example of how variation in the quality and the amount of food could impact on longevity by modulating signals that are influenced by genetic variants selected in centenarians. To be noted, Sirtuin 1 (SIRT1), despite its role in many cellular processes, induces survival and is not affected by polymorphisms that associate with exceptional longevity in humans, perhaps on account of its critical role in tumours as well [18].

The candidate gene approach and associations with longevity

To date, only a few genetic variants have been consistently found associated with exceptional longevity in humans (Table 1). The most convincing result discovered to date by a candidate gene approach is the reduction of the APOE 4 allele in centenarians as a result of its correlation with cardiovascular disease and Alzheimer's disease [7]. APOE knockout mice develop atherosclerosis, so genes that modulate vascular integrity are potential candidates that may be identified in genetic association studies of exceptional longevity [19].

The FOXO3A rs2802292 allele is another variant found associated with exceptional longevity across populations [20,22-24]. Nevertheless, this polymorphism has no apparent impact on the functions of FOXO3A and it is not in LD with functional variants. However, FOXO3A is part of a pathway associated with longevity: IGF1/PI3K/PDK1/AKT/FOXO. Animal studies from worms to mice have shown that genetic modifications are able to postpone aging by modulating the IGF1/PI3K/PDK1/AKT/FOXO pathway [25]. This pathway regulates many aspects of cell homeostasis, from cell survival and proliferation to oxidative stress response, depending on concomitant stimuli [26,27]. Interestingly, individuals with short stature due to a lack of growth hormone, which is upstream of the IGF1/PI3K/PDK1/AKT/FOXO pathway, have a reduced incidence of tumours and diabetes [28]. The IGF1/PI3K/PDK1/AKT/FOXO pathway is strongly modulated by caloric restriction, as are the levels of AMPK and SIRT1. Glucose and insulin resistance also modulates this pathway, which associates with inefficient uptake of glucose, inducing its overstimulation and aging. To some extent, genetic alterations of the insulin-like growth factor 1 (IGF1) receptor that alter the IGF signalling pathway confer an increase in propensity for longevity [29].

Adenosine deaminase, RNA-specific (ADAR) and telomerase gene variants have also been associated with human longevity [30,31]. However, with the exception of APOE and FOXO3A variants, none of the many candidate genetic variants tested to date have been consistently replicated across populations. This is possibly on account of differing environmental stimuli generating inconsistent demographic pressures, making results, as a consequence, irreproducible [32].

Other potential problems that generate false positive and false negative results include the low power of studies using small sample sizes, and a lack of a suitable control for the genetic admixture. On the first point, an

Table 1 Genes and variants found correlated with longevity in humans

Gene	Variant	Occurrence in centenarians	Previous disease correlations	Potential role in longevity	Ref.
APOE	ε4*	reduced	Alzheimer's and cardiovascular diseases	Maintenance of vascular integrity	[7]
FOXO3A	rs2802292*	more present	none	Control of cell homeostasis	[20]
CAMK4	rs10491334[†]	more present	hypertension	Modulation of CREB, SIRT and FOXO3A	[21]
ATXN1	rs697739	more present	amyotrophic lateral sclerosis (age of onset)	Modulation of CREB	[21]
DCAMKL1	rs9315385	more present	heart rate variability	Modulation of CREB	[21]

*, result obtained by a candidate gene approach and replicated in other studies; [†], result obtained from a GWAS and confirmed in a replication cohort.

extremely instructive review has been written by Altshuler, Daly and Lender, who calculate the power of a study based on the number of individuals genotyped, the number of tested hypotheses, and the frequency of the allele tested for a specific OR [33]. From the graph given in their review (Figure 1), it is clear that for the OR expected in human exceptional longevity (between 1.2 and 2), the power of a study is highly dependent upon the number of hypotheses tested. If we consider that many laboratories test their genetic variants and publish only the positive results they find, the few hundred individuals typically used in a candidate gene approach on exceptional longevity are not sufficient to minimize false-positive findings.

With regard to the appropriate control for a given genetic admixture, it is possible to correct for systematic ancestry differences between cases and controls — an effect that can cause false associations — by the application of principal

components analysis to the genotyping of thousands of SNPs with chips [34].

Finally, because the candidate gene approach is hypothesis driven, the functional validation of the coded protein in signals important for aging and longevity do not add strength to the finding, while as we will see, this is an opportunity for hypothesis-free approaches, such as the genome-wide association study (GWAS) and sib-pair analysis.

Genome-wide association studies and exceptional longevity

GWASs are hypothesis-free efforts that generate findings that need to be replicated in independent populations. In the case of exceptional longevity, success in replicating initial findings is negatively influenced by the differences in participant ages, gender and disease status distribution across the analysed populations. Furthermore, the

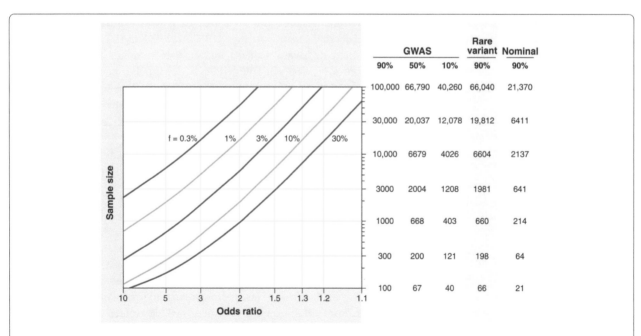

Figure 1 Sample sizes required for genetic association studies. The graph shows the total number N of samples (consisting of N/2 cases and N/2 controls) required to map a genetic variant as a function of the increased risk due to the disease-causing allele (x axis) and the frequency of the disease-causing allele (various curves). The required sample size is shown in the table on the right for various different kinds of association studies [33]. Reproduced with permission from The American Association for the Advancement of Science.

GWAS approach suffers from the multiple-testing statistical penalty that forces the adoption of very low p values of significance, hence favouring the phenomenon of the winning course, i.e. the enrichment of false-positive associations among the dozens of top findings [35]. For these reasons, recent GWASs have failed to find variants that cross-validate across populations — with the exception of the known APOE locus — pointing to the need of much larger studies or alternative study designs in order to discover common polymorphisms with smaller genetic effects and rare variants with high penetrance that influence exceptional longevity [36,37].

Regarding the power of capturing true associations in GWAS efforts, the Altshuler, Lander and Daly review calculation clearly shows that a sample size comprising thousands of individuals is needed to identify the expected OR in a GWAS on exceptional longevity [33]. Thus, a GWAS on exceptional longevity can be considered only a hypothesis-generating effort to be used in conjunction with other studies.

For the above reasons, in our recently published GWAS on individuals enrolled in the Southern Italian Centenarian Study (SICS), we not only attempted to progressively reduce the number of tested hypotheses, but also considering the redundancy (non-independence) of the many SNPs represented on the Illumina 317 k beadchip employed, we decided to use a 300 k SNP screening of the SICS individuals as a hypothesis-generating set, adopting a Genomic Control (GC)-corrected p-value $< 1e-4$ threshold (which is a less stringent threshold than $p < 5 \times 10\text{-}2/317000 = 1.5 \times 10\text{-}7$) for the replication, evaluating allelic, genotypic, dominant and recessive genetic association models [21].

Initial screening of SICS individuals identified CAMK4 rs10491334, a variant that had been already established among the top 5 SNPs in the Framingham Heart Study on diastolic high blood pressure [38]. The fact that CAMK4 rs10491334 associates also with hypertension is reassuring in that hypertension and longevity are regulated by common pathways. In fact, mice with genetic ablation of the angiotensin II type1 receptor — the key regulator of blood pressure — had increased expression of the longevity gene Sirt3 and improved survival [39]. Interestingly, rs10491334 correlated with CAMK4 protein expression, and functional studies revealed the ability of CAMK4 protein to modulate SIRT1 and FOXO3A.

The ataxin-1 (ATXN1) rs697739 allele was another variant found among the top findings of our GWAS on SICS individuals. This polymorphism had been previously associated with the age of onset of sporadic amyotrophic lateral sclerosis, a disease of unknown cause characterized by slowly progressive degeneration of motor neurons and that usually occurs in patients aged 40–60 years [40]. ATXN1 is the gene responsible for spinocerebellar ataxia type 1 and

antagonizes the neuronal survival function of myocyte enhancer factor-2 (MEF2) [41].

MEF2 transcription repression by cabin1-HDAC4 is removed by CAMKIV activation, and this suggests that MEF2 is a common downstream target of CAMKIV and ATXN1 [42,43].

In addition to CAMKIV rs10491334 and ATXN1 rs697739, the rs9315385 allele of doublecortin and Ca^{2+}/calmodulin-dependent kinase-like-1 (DCAMKL1) was a third top finding of our study. DCAMKL1 has structural similarity with CAMKIV, but despite this, it represses CAMKIV-induced activation of cAMP response element-binding (CREB) protein via phosphorylation of transducer of regulated CREB activity 2 (TORC2) at Ser171 [44]. DCAMKL1 rs9315385 was previously associated with total power of HRV [45]. A reduced HRV is a marker of autonomic dysfunction and is associated with an increased risk of cardiovascular morbidity and mortality [46]. HRV-parasympathetic function decreases up to the eighth decade of life, followed by an increase to higher levels — similar to those found in a younger population — in nonagenarians and centenarians [13]. Similarly to CAMKIV, DCAMKL1 and ATXN1 are expressed mainly in brain. These data support the importance of the CAMKIV/CREB pathway in regulating the aging process.

A brief mention needs to be made here on the cutting-edge, genetic signature paper by Sebastiani et al. that very elegantly proved that a complex analysis on 281 SNPs allowed to define clusters of individuals that aged differently based on their genetic signature [47].

Linkage analysis and exome sequencing

Sib-pair analysis has been for a while the only tool available for the identification of chromosomal regions that potentially harbour genetic variants influencing the phenotype of interest. The approach can identify excess allele sharing, and was initially performed with microsatellites. It consists in the identical-by-descendant analysis of very informative markers that reconstruct the haplotype of parents and how they co-segregate in their offspring. We performed such an analysis on a unique collection of sib-pairs and their families, collected by the New England Centenarian Study (NECS), and identified a significant peak on 4q25 [48]. Follow-up analysis failed to identify genetic variants that could explain the initial linkage finding. Rare mutations that segregate in centenarian siblings are eventually captured by sibling-pair analysis, but this cannot be the case for genetic association studies that loose power as the allele frequency of the tested polymorphisms drops. Furthermore, it is possible that linkage efforts identify chromosomal regions where more causative genetic variants reside, and thus the sum of their effects determines the linkage result, whereas with follow-up genetic association approaches, the analysis involves one common polymorphism at a time or, eventually, haplotypes. Attempts to replicate the initial linkage did not succeed, except for an initial replication effort that successfully replicated the linkage at D4S1564 [49,50]. A negative replication effort can be due to an initial false-positive finding or to the diversity of the populations used for the replication effort, in terms of genetic background, the environment applying the demographic pressure, the ages of the participants, the number of families and the genetic markers adopted. Recently, Kunkel's laboratory published a well-performed re-analysis of a part of the sib-pairs used in the initial study, plus new sib-pairs recruited by Elixir Pharmaceuticals [51]. To be noted, some of the largest and more impressive families — those that were genotyped upfront and that showed immediately a significant linkage on 4q25 in the original study — were either not analysed or done so only in part by this second effort. The new analysis adopted a high-density marker panel of SNPs to genotype the patients, allowing a better coverage of the genome. They did not replicate the chromosome 4q25 finding, except when the same stringent criteria were adopted to select a sub-set of centenarian families. Interestingly, a new peak on chromosome 3p24-22 reached significant threshold, and a second peak was highly suggestive of linkage at 9q31-34. This latter peak appeared also in the previous analysis with microsatellites, even if less robustly. The attempt to identify the genetic variant/variants responsible for the 4q25 peak pointed to the initial encouraging genetic variant in the promoter of the microsomal triglyceride transfer protein (MTP) gene [52]. Unfortunately, the finding was not replicated by an independent effort and by our analysis that included more controls [32,53].

It is plausible that different approaches are needed to follow up genetic linkage findings, to point to the identification of rare variants that co-segregate in families. To this end, exome sequencing data, intersected with linkage data, could give rise to interesting results.

To be noted, the 4q25 locus harbours elongation of very long chain fatty acids protein 6 (ELOVL6), the elongase that transforms C16:0 into C18:0 and C16:1 into C18:1. Polymorphisms in this gene have been associated with insulin sensitivity; a mouse deficient for this gene carried high doses of C16:1 (palmitoleic acid) and did not acquire insulin resistance after a high-fat diet [54,55]. C16:1 has been identified as an adipose tissue-derived lipid hormone that strongly stimulates muscle insulin action and suppresses hepatosteatosis [56]. Genetically modified, long-living worms have an incredible correlation between their increase in life-span and their palmitoleic acids levels [57]. This is stunning if we consider the increased level of palmitoleic acid that we observed in centenarians' offspring and that the gene of the major modifier of palmitoleic acid levels (i.e. ELOVL6) is located in the 4q25 longevity locus [11,48]. The re-sequencing in centenarians of this gene could bring to the identification of rare variants able to influence its activity.

Thus, the old approach of linkage analysis when combined with the new technologies of high massive re-sequencing could produce novel and interpretable results. Re-sequencing alone, because of the enormous amount of information generated, would force the application of a huge statistical correction for the multiple testing, which would cause the loss of most, if not all, the potential findings, as happens with GWAS.

Furthermore, multivariate models, based on machine-learning algorithms (i.e. Bayesian networks [58], classification and regression trees — CART [59] — and support vector machines — SVM [60]), are able to overcome the limitations of the usual *"one-SNP-at-the-time"* testing strategies usually employed for identifying causative variants. In particular, these kinds of approaches allow for a more in-depth comprehension of the molecular mechanisms underlying multifactorial traits, such as longevity, which result from the interaction of genetic variants (SNPs, mutations) and environmental and clinical determinants (e.g. diet, stress, comorbidities). In this context, bioinformatics plays a key role, allowing genetic information to be managed at a genome-wide level and to be integrated with the clinical information available.

Concluding remarks

Despite the enormous progress achieved by DNA-investigating technologies, such as SNP arrays and exome capturing/re-sequencing, the current knowledge on how genetic variants influence exceptional longevity in humans is still based on the old candidate gene approaches. The adoption of innovative study designs combined with novel genetic platforms and innovative statistical methods hopefully will bring to the identification of new intervention points at which to modulate aging and the diseases of aging.

Abbreviations
ADAR: adenosine deaminase RNA-specific; AMPK: Adenosine monophosphate-activated protein kinase; APOE: apolipoprotein E; ATXN1: Ataxin 1; CAMK: Ca^{2+}/calmodulin-dependent protein kinase; CART: classification and regression tree; CREB: cAMP response element-binding; DCAMKL1: doublecortin and Ca^{2+}/calmodulin-dependent kinase-like-1; ELOVL6: elongation of very long chain fatty acids protein 6; EPIC: Elderly Prospective Cohort Study; FOXO: forkhead box O; GC: genomic control; GWAS: genome-wide association study; HRV: heart-rate variability; HSP: heat shock protein; IGF: insulin-like growth factor; LD: linkage disequilibrium; MEF2: myocyte enhancer factor-2; MTP: microsomal triglyceride transfer protein; NECS: New England Centenarian Study; OR: odds ratio; SICS: Southern Italian Centenarian Study; SIRT1: Sirtuin 1; SNP: single nucleotide polymorphism; SVM: support vector machine; TORC2: transducer of regulated CREB activity 2.

Competing interests
The authors declare that they have no competing interests.

Acknowledgements
Anna Ferrario was supported by a Fondazione Umberto Veronesi fellowship. We thank Michael V. G. Latronico for careful revision of the manuscript.

Author details
[1]IRCCS Multimedica, Via Fantoli 16/15, 20138 Milan, Italy. [2]University of Pavia, Via Ferrata 1, 27100 Pavia, Italy. [3]Università degli Studi di Salerno, Via S. Allende, 84081 Baronissi, Italy.

Authors' contributions
AAP wrote the first draft; Subsequent drafts were written by AF, FV and FA, who had the overall supervision of the review processing; all authors edited the paper and approved its final version.

Received: 27 March 2012 Accepted: 23 April 2012
Published: 23 April 2012

References
1. Perls TT: **The different paths to 100.** *Am J Clin Nutr* 2006, **83**:484S–487S.
2. Oeppen J, Vaupel JW: **Demography. Broken limits to life expectancy.** *Science* 2002, **296**:1029–1031.
3. Trichopoulou A, Orfanos P, Norat T, Bueno-de-Mesquita B, Ocke MC, Peeters PH, van der Schouw YT, Boeing H, Hoffmann K, Boffetta P, *et al*: **Modified Mediterranean diet and survival: EPIC-elderly prospective cohort study.** *BMJ* 2005, **330**:991.
4. Terry DF, Sebastiani P, Andersen SL, Perls TT: **Disentangling the roles of disability and morbidity in survival to exceptional old age.** *Arch Intern Med* 2008, **168**:277–283.
5. Perls T, Shea-Drinkwater M, Bowen-Flynn J, Ridge SB, Kang S, Joyce E, Daly M, Brewster SJ, Kunkel L, Puca AA: **Exceptional familial clustering for extreme longevity in humans.** *J Am Geriatr Soc* 2000, **48**:1483–1485.
6. Perls TT, Wilmoth J, Levenson R, Drinkwater M, Cohen M, Bogan H, Joyce E, Brewster S, Kunkel L, Puca A: **Life-long sustained mortality advantage of siblings of centenarians.** *Proc Natl Acad Sci U S A* 2002, **99**:8442–8447.
7. Schachter F, Faure-Delanef L, Guenot F, Rouger H, Froguel P, Lesueur-Ginot L, Cohen D: **Genetic associations with human longevity at the APOE and ACE loci.** *Nat Genet* 1994, **6**:29–32.
8. Terry DF, Wilcox MA, McCormick MA, Pennington JY, Schoenhofen EA, Andersen SL, Perls TT: **Lower all-cause, cardiovascular, and cancer mortality in centenarians' offspring.** *J Am Geriatr Soc* 2004, **52**:2074–2076.
9. Terry DF, McCormick M, Andersen S, Pennington J, Schoenhofen E, Palaima E, Bausero M, Ogawa K, Perls TT, Asea A: **Cardiovascular disease delay in centenarian offspring: role of heat shock proteins.** *Ann N Y Acad Sci* 2004, **1019**:502–505.
10. Barzilai N, Atzmon G, Schechter C, Schaefer EJ, Cupples AL, Lipton R, Cheng S, Shuldiner AR: **Unique lipoprotein phenotype and genotype associated with exceptional longevity.** *JAMA* 2003, **290**:2030–2040.
11. Puca AA, Andrew P, Novelli V, Anselmi CV, Somalvico F, Cirillo NA, Chatgilialoglu C, Ferreri C: **Fatty acid profile of erythrocyte membranes as possible biomarker of longevity.** *Rejuvenation Res* 2008, **11**:63–72.
12. Paolisso G, Gambardella A, Ammendola S, D'Amore A, Balbi V, Varricchio M, D'Onofrio F: **Glucose tolerance and insulin action in healty centenarians.** *Am J Physiol* 1996, **270**:E890–E894.
13. Zulfiqar U, Jurivich DA, Gao W, Singer DH: **Relation of high heart rate variability to healthy longevity.** *Am J Cardiol* 2010, **105**:1181–1185.
14. Hardie DG: **Adenosine Monophosphate-Activated Protein Kinase: A Central Regulator of Metabolism with Roles in Diabetes, Cancer, and Viral Infection.** *Cold Spring Harb Symp Quant Biol* 2011, [Epub ahead of print].
15. Zong H, Ren JM, Young LH, Pypaert M, Mu J, Birnbaum MJ, Shulman GI: **AMP kinase is required for mitochondrial biogenesis in skeletal muscle in response to chronic energy deprivation.** *Proc Natl Acad Sci U S A* 2002, **99**:15983–15987.
16. Hardie DG: **AMPK and autophagy get connected.** *EMBO J* 2011, **30**:634–635.
17. Hardie DG: **AMPK: a key regulator of energy balance in the single cell and the whole organism.** *Int J Obes (Lond)* 2008, **32**(Suppl 4):S7–S12.
18. Flachsbart F, Croucher PJ, Nikolaus S, Hampe J, Cordes C, Schreiber S, Nebel A: **Sirtuin 1 (SIRT1) sequence variation is not associated with exceptional human longevity.** *Exp Gerontol* 2006, **41**:98–102.
19. Imaizumi K: **Diet and atherosclerosis in apolipoprotein E-deficient mice.** *Biosci Biotechnol Biochem* 2011, **75**:1023–1035.
20. Willcox BJ, Donlon TA, He Q, Chen R, Grove JS, Yano K, Masaki KH, Willcox DC, Rodriguez B, Curb JD: **FOXO3A genotype is strongly associated with human longevity.** *Proc Natl Acad Sci U S A* 2008, **105**:13987–13992.
21. Malovini A, Illario M, Iaccarino G, Villa F, Ferrario A, Roncarati R, Anselmi CV, Novelli V, Cipolletta E, Leggiero E, et al: **Association study on long-living individuals from Southern Italy identifies rs10491334 in the CAMKIV gene that regulates survival proteins.** *Rejuvenation Res* 2011, **14**:283–291.
22. Anselmi CV, Malovini A, Roncarati R, Novelli V, Villa F, Condorelli G, Bellazzi R, Puca AA: **Association of the FOXO3A locus with extreme longevity in a southern Italian centenarian study.** *Rejuvenation Res* 2009, **12**:95–104.
23. Flachsbart F, Caliebe A, Kleindorp R, Blanche H, von Eller-Eberstein H, Nikolaus S, Schreiber S, Nebel A: **Association of FOXO3A variation with human longevity confirmed in German centenarians.** *Proc Natl Acad Sci U S A* 2009, **106**:2700–2705.
24. Soerensen M, Dato S, Christensen K, McGue M, Stevnsner T, Bohr VA, Christiansen L: **Replication of an association of variation in the FOXO3A gene with human longevity using both case–control and longitudinal data.** *Aging Cell* 2010, **9**:1010–1017.
25. Kenyon CJ: **The genetics of ageing.** *Nature* 2010, **464**:504–512.
26. Kops GJ, Dansen TB, Polderman PE, Saarloos I, Wirtz KW, Coffer PJ, Huang TT, Bos JL, Medema RH, Burgering BM: **Forkhead transcription factor FOXO3a protects quiescent cells from oxidative stress.** *Nature* 2002, **419**:316–321.
27. van der Horst A, Burgering BM: **Stressing the role of FoxO proteins in lifespan and disease.** *Nat Rev Mol Cell Biol* 2007, **8**:440–450.
28. Guevara-Aguirre J, Balasubramanian P, Guevara-Aguirre M, Wei M, Madia F, Cheng CW, Hwang D, Martin-Montalvo A, Saavedra J, Ingles S, et al: **Growth hormone receptor deficiency is associated with a major reduction in pro-aging signaling, cancer, and diabetes in humans.** *Sci Transl Med* 2011, **3**:70ra13.
29. Pawlikowska L, Hu D, Huntsman S, Sung A, Chu C, Chen J, Joyner AH, Schork NJ, Hsueh WC, Reiner AP, *et al*: **Association of common genetic variation in the insulin/IGF1 signaling pathway with human longevity.** *Aging Cell* 2009, **8**:460–472.
30. Sebastiani P, Montano M, Puca A, Solovieff N, Kojima T, Wang MC, Melista E, Meltzer M, Fischer SE, Andersen S, *et al*: **RNA editing genes associated with extreme old age in humans and with lifespan in C. elegans.** *PLoS One* 2009, **4**:e8210.
31. Atzmon G, Cho M, Cawthon RM, Budagov T, Katz M, Yang X, Siegel G, Bergman A, Huffman DM, Schechter CB, et al: **Evolution in health and medicine Sackler colloquium: Genetic variation in human telomerase is associated with telomere length in Ashkenazi centenarians.** *Proc Natl Acad Sci U S A* 2010, **107** (Suppl 1):1710–1717.
32. Novelli V, Viviani Anselmi C, Roncarati R, Guffanti G, Malovini A, Piluso G, Puca AA: **Lack of replication of genetic associations with human longevity.** *Biogerontology* 2008, **9**:85–92.
33. Altshuler D, Daly MJ, Lander ES: **Genetic mapping in human disease.** *Science* 2008, **322**:881–888.
34. Price AL, Patterson NJ, Plenge RM, Weinblatt ME, Shadick NA, Reich D: **Principal components analysis corrects for stratification in genome-wide association studies.** *Nat Genet* 2006, **38**:904–909.
35. Ioannidis JP, Thomas G, Daly MJ: **Validating, augmenting and refining genome-wide association signals.** *Nat Rev Genet* 2009, **10**:318–329.
36. Deelen J, Beekman M, Uh HW, Helmer Q, Kuningas M, Christiansen L, Kremer D, van der Breggen R, Suchiman HE, Lakenberg N, et al: **Genome-wide association study identifies a single major locus contributing to survival into old age; the APOE locus revisited.** *Aging Cell* 2011, **10**:686–698.
37. Nebel A, Kleindorp R, Caliebe A, Nothnagel M, Blanche H, Junge O, Wittig M, Ellinghaus D, Flachsbart F, Wichmann HE, et al: **A genome-wide association study confirms APOE as the major gene influencing survival in long-lived individuals.** *Mech Ageing Dev* 2011, **132**:324–330.
38. Levy D, Larson MG, Benjamin EJ, Newton-Cheh C, Wang TJ, Hwang SJ, Vasan RS, Mitchell GF: **Framingham Heart Study 100 K Project: genome-wide associations for blood pressure and arterial stiffness.** *BMC Med Genet* 2007, **8**(Suppl 1):S3.

39. Benigni A, Corna D, Zoja C, Sonzogni A, Latini R, Salio M, Conti S, Rottoli D, Longaretti L, Cassis P, *et al*: Disruption of the Ang II type 1 receptor promotes longevity in mice. *J Clin Invest* 2009, **119**:524–530.

40. Landers JE, Melki J, Meininger V, Glass JD, van den Berg LH, van Es MA, Sapp PC, van Vught PW, McKenna-Yasek DM, Blauw HM, *et al*: Reduced expression of the Kinesin-Associated Protein 3 (KIFAP3) gene increases survival in sporadic amyotrophic lateral sclerosis. *Proc Natl Acad Sci U S A* 2009, **106**:9004–9009.

41. Bolger TA, Zhao X, Cohen TJ, Tsai CC, Yao TP: The neurodegenerative disease protein ataxin-1 antagonizes the neuronal survival function of myocyte enhancer factor-2. *J Biol Chem* 2007, **282**:29186–29192.

42. Blaeser F, Ho N, Prywes R, Chatila TA: Ca(2+)-dependent gene expression mediated by MEF2 transcription factors. *J Biol Chem* 2000, **275**:197–209.

43. Racioppi L, Means AR: Calcium/calmodulin-dependent kinase IV in immune and inflammatory responses: novel routes for an ancient traveller. *Trends Immunol* 2008, **29**:600–607.

44. Ohmae S, Takemoto-Kimura S, Okamura M, Adachi-Morishima A, Nonaka M, Fuse T, Kida S, Tanji M, Furuyashiki T, Arakawa Y, *et al*: Molecular identification and characterization of a family of kinases with homology to Ca2+/calmodulin-dependent protein kinases I/IV. *J Biol Chem* 2006, **281**:20427–20439.

45. Newton-Cheh C, Guo CY, Wang TJ, O'Donnell CJ, Levy D, Larson MG: Genome-wide association study of electrocardiographic and heart rate variability traits: the Framingham Heart Study. *BMC Med Genet* 2007, **8** (Suppl 1):S7.

46. Thayer JF, Yamamoto SS, Brosschot JF: The relationship of autonomic imbalance, heart rate variability and cardiovascular disease risk factors. *Int J Cardiol* 2010, **141**:122–131.

47. Sebastiani P, Solovieff N, Dewan AT, Walsh KM, Puca A, Hartley SW, Melista E, Andersen S, Dworkis DA, Wilk JB, et al: Genetic signatures of exceptional longevity in humans. *PLoS One* 2012, **7**:e29848.

48. Puca AA, Daly MJ, Brewster SJ, Matise TC, Barrett J, Shea-Drinkwater M, Kang S, Joyce E, Nicoli J, Benson E, *et al*: A genome-wide scan for linkage to human exceptional longevity identifies a locus on chromosome 4. *Proc Natl Acad Sci U S A* 2001, **98**:10505–10508.

49. Reed T, Dick DM, Uniacke SK, Foroud T, Nichols WC: Genome-wide scan for a healthy aging phenotype provides support for a locus near D4S1564 promoting healthy aging. *J Gerontol A Biol Sci Med Sci* 2004, **59**:227–232.

50. Beekman M, Blauw GJ, Houwing-Duistermaat JJ, Brandt BW, Westendorp RG, Slagboom PE: Chromosome 4q25, microsomal transfer protein gene, and human longevity: novel data and a meta-analysis of association studies. *J Gerontol A Biol Sci Med Sci* 2006, **61**:355–362.

51. Boyden SE, Kunkel LM: High-density genomewide linkage analysis of exceptional human longevity identifies multiple novel loci. *PLoS One* 2010, **5**:e12432.

52. Geesaman BJ, Benson E, Brewster SJ, Kunkel LM, Blanche H, Thomas G, Perls TT, Daly MJ, Puca AA: Haplotype-based identification of a microsomal transfer protein marker associated with the human lifespan. *Proc Natl Acad Sci U S A* 2003, **100**:14115–14120.

53. Nebel A, Croucher PJ, Stiegeler R, Nikolaus S, Krawczak M, Schreiber S: No association between microsomal triglyceride transfer protein (MTP) haplotype and longevity in humans. *Proc Natl Acad Sci U S A* 2005, **102**:7906–7909.

54. Matsuzaka T, Shimano H, Yahagi N, Kato T, Atsumi A, Yamamoto T, Inoue N, Ishikawa M, Okada S, Ishigaki N, *et al*: Crucial role of a long-chain fatty acid elongase, Elovl6, in obesity-induced insulin resistance. *Nat Med* 2007, **13**:1193–1202.

55. Morcillo S, Martin-Nunez GM, Rojo-Martinez G, Almaraz MC, Garcia-Escobar E, Mansego ML, de Marco G, Chaves FJ, Soriguer F: ELOVL6 genetic variation is related to insulin sensitivity: a new candidate gene in energy metabolism. *PLoS One* 2011, **6**:e21198.

56. Cao H, Gerhold K, Mayers JR, Wiest MM, Watkins SM, Hotamisligil GS: Identification of a lipokine, a lipid hormone linking adipose tissue to systemic metabolism. *Cell* 2008, **134**:933–944.

57. Shmookler Reis RJ, Xu L, Lee H, Chae M, Thaden JJ, Bharill P, Tazearslan C, Siegel E, Alla R, Zimniak P, Ayyadevara S: Modulation of lipid biosynthesis contributes to stress resistance and longevity of C. elegans mutants. *Aging (Albany NY)* 2011, **3**:125–147.

58. Friedman N, Geiger D, Goldszmidt M: Bayesian Network Classifiers. *Mach Learn* 1997, **29**:131–163.

59. Breiman L, Friedman JH, Olshen RA, Stone CJ: *Classification and regression trees*. Monterey: Wadsworth & Brooks/Cole Advanced Books & Software; 1984.

60. Crammer K, Singer Y: On the algorithmic implementation of multiclass kernel-based vector machines. *J Mach Learn Res* 2001, **2**:265–292.

doi:10.1186/1742-4933-9-7
Cite this article as: Ferrario *et al.*: The application of genetics approaches to the study of exceptional longevity in humans: potential and limitations. *Immunity & Ageing* 2012 **9**:7.

Balistreri *et al. Immunity & Ageing* 2012, **9**:8
http://www.immunityageing.com/content/9/1/8

IMMUNITY & AGEING

RESEARCH   **Open Access**

Genetics of longevity. Data from the studies on Sicilian centenarians

Carmela R Balistreri[1*], Giuseppina Candore[1], Giulia Accardi[1], Manuela Bova[1], Silvio Buffa[1], Matteo Bulati[1], Giusi I Forte[1], Florinda Listì[1], Adriana Martorana[1], Marisa Palmeri[1], MariaValeria Pellicanò[1,2], Loredana Vaccarino[1], Letizia Scola[1], Domenico Lio[1] and Giuseppina Colonna-Romano[1]

Abstract

The demographic and social changes of the past decades have determined improvements in public health and longevity. So, the number of centenarians is increasing as a worldwide phenomenon. Scientists have focused their attention on centenarians as optimal model to address the biological mechanisms of "successful and unsuccessful ageing". They are equipped to reach the extreme limits of human life span and, most importantly, to show relatively good health, being able to perform their routine daily life and to escape fatal age-related diseases, such as cardiovascular diseases and cancer. Thus, particular attention has been centered on their genetic background and immune system. In this review, we report our data gathered for over 10 years in Sicilian centenarians. Based on results obtained, we suggest longevity as the result of an optimal performance of immune system and an over-expression of anti-inflammatory sequence variants of immune/inflammatory genes. However, as well known, genetic, epigenetic, stochastic and environmental factors seem to have a crucial role in ageing and longevity. Epigenetics is associated with ageing, as demonstrated in many studies. In particular, ageing is associated with a global loss of methylation state. Thus, the aim of future studies will be to analyze the weight of epigenetic changes in ageing and longevity.

Keywords: Immune system, Genetics, Pro/anti-inflammatory polymorphisms, Epigenomics

Introduction

Data from centenarian offspring

As well known, life expectancy is a familial trait and longevity is determined by different factors. In particular, the environmental milieu and genetic background play a central role. As demonstrated by many epidemiological studies, family members of long-lived subjects have a significant survival advantage compared to general population. In this context, the study of centenarian offspring (CO), a group of healthy elderly people with a familiar history of longevity, might help gerontologists to better identify the correlation between genetic profile and hope of a healthy ageing. Previous studies have reported that CO, like their centenarian parents, have genetic and immune system advantages, which reflect a minor risk to develop major age-related diseases, such

as cardiovascular diseases, hypertension or diabetes mellitus as well as cancer [1,2]. The lower cardiovascular disease risk in CO suggests the probability that CO have some protective factors against atherosclerosis, such as a good lipid profile. Male CO have higher plasma HDL-C levels and lower plasma LDL-C levels. Since lipid profile is directly correlated to atherosclerotic cardiovascular diseases, this metabolic feature could preserve CO both to develop these diseases and, as consequence, to reach a healthy ageing and longer survival [3]. Furthermore, Rose et al. [4] reported that centenarians and CO show significantly higher levels of heteroplasmy in mtDNA control region than controls, a favorable condition for longevity.

In these last years, some researchers have speculated about the distinctive immunological profile of offspring enriched for longevity respect to the immunological features of coeval elderly. The cytomegalovirus (CMV) is one of the most common viruses that affect elderly people. Many evidences have shown that CMV infection

* Correspondence: carmelarita.balistreri@unipa.it
[1]Department of Pathobiology and Medical and Forensic Biotechnologies, University of Palermo, Corso Tukory 211, Palermo 90134, Italy
Full list of author information is available at the end of the article

may influence the T cell subset distribution, having an essential role in immunosenescence [5-7]. CMV infection is strongly related to both a reduction of CD8$^+$CD45$^+$CCR7$^+$CD27$^+$CD28$^+$ naïve T cells and to a contemporarily increase of CD8$^+$CD45RA$^-$CCR7$^-$CD27$^-$CD28$^-$ late differentiated effector memory and CD45RA re-expressing T cells. These parameters are considered typical of immunosenescence in elderly. Recently, it has been demonstrated that CMV-seropositive offspring of long-lived people don't show the age-associated decrease of naïve T cells. On the other hand, memory T cell subsets above described do not increase in offspring of long-lived families, differently from that observed in age-matched controls [8]. It has been also demonstrated that CMV-seropositive offspring of long-lived people have reduced levels of CD8+ T cells expressing CD57 and KLRG1, sometimes referred as "marker of senescence", when compared to their CMV-infected age-matched controls. The reduction of effector memory T cells lacking the expression of CD27 and CD28 and expressing CD57 and KLRG1, observed in CMV-infected offspring could explain their high proliferative response against CMV. The CMV-seropositive offspring have also shown significantly lower CRP levels compared to their CMV-seropositive age-matched controls that could be related to a lower pro-inflammatory status [8].

During ageing, B cell compartment also shows significant modifications in numbers and functions [9-12]. In fact, advanced age is per se a condition characterized by lack of B clonotypic immune response to new extracellular pathogens. In any event, data are suggesting that the loss of naive B cells could represent a hallmark of immunosenescence [13]. On the other hand, a B cell population lacking of both IgD and CD27 resulted increased in healthy elderly [14]. We have suggested that this IgD$^-$CD27$^-$ B cell subset is a population of memory B cells lacking CD27, a typical memory marker, likely considered a late memory exhausted B cell subset (Table 1) [14-16]. This population resulted also increased in active

Table 1 Main modifications of B cells and B cells products in elderly human observed in our laboratory

B cells or B cells products	Changes	References
Total B cells (percentage)	↓	[9]
CD19$^+$CD5$^+$ B1 cells (percentage and absolute number)	↓	[10]
IgG, IgA	↑	[11]
IgM, IgD	↓	[11]
IgE	=	[11]
Autoantibodies	↑	[12]
Naive (IgD$^+$CD27$^-$)	↓	[13]
DN (IgD$^-$CD27$^-$)	↑	[14-16]

Lupus patients [17], in healthy subjects challenged with respiratory syncitial virus [18], and in HIV patients [19]. CO don't show the typical naïve/memory B cell shift observed in elderly. Although a decreased B cell count was observed in CO and their age-matched controls, it has been demonstrated that naïve B cells (IgD$^+$CD27$^-$) were more abundant and DN B cells (IgD$^-$CD27$^-$) were significantly decreased, as looked similarly in young people [20]. This B cells distribution in CO could suggest that antigenic load or inflammatory environment play a central role in exhaustion of the B cell branch. It is well documented that the quality and the size of the humoral immune response declines with age [15,21-26]. This change is characterized by lower antibody responses and decreased production of high affinity antibodies. The evaluation of IgM secreted in CO serum shows that the values are within the range of the levels observed in young subjects [20]. In this way, CO could have a bigger advantage to fight against new infections and appropriately respond to vaccinations, giving them a selective advantage for longevity in healthiness.

In conclusion, individuals genetically enriched for longevity possess immune different signatures respect to those of the general population (Table 2). This suggests the idea of the "familiar youth" of the immune system. In addition, the lower pro-inflammatory status in CMV-infected offspring of long-lived people might represent an optimal advantage for healthy longevity and against mortality associated to major age-related diseases.

Gender and longevity

A characteristic enigma of longevity is the gender and the social phenomenon of *"feminization of old age"*. The demographic and social changes of the past decades, responsible for longevity and the improvements in public health, have created new and often very dissimilar realties for women and men. People are all aware that they differ in their anatomy and physiology, but also in more complex traits, such as lifespan (in Italy, 78.8 years for men and 84.1 years for women, respectively) and mortality [27-29]. No conclusive explanation for these new differences is actually demonstrated. An intricate interaction between environmental, social structural, behavioural (i.e. the complex pattern of roles and values that define what is thought as *masculine* and *feminine*) and genetic factors have been suggested as the more probable reason [30-32].

From a genetic prospective, our suggestion based on the studies in Sicilian population supports a female-specific gene-longevity association, by emphasizing the paradoxical role of socio-cultural habits in female longevity [33]. This concerns the HFE gene, the most telomeric HLA class I gene, codifying for a class I α chain, the HFE protein, which seemingly no longer participates

Table 2 Cellular and humoral immune modification in offspring from longevity families compared to their AM controls

T and B cell Phenotypes and Products	Changes	References
Naïve T cells (CD3$^+$CD8$^+$CD45RA$^+$CCR7$^+$CD27$^+$CD28$^+$)	Increase	[8]
Late differentiated effector memory T cells (CD3$^+$CD8$^+$CD45RA$^-$CCR7$^-$CD27$^-$CD28$^-$)	Decrease	[8]
TEMRA (CD3$^+$CD8$^+$CD45RA$^+$CCR7$^-$CD27$^-$CD28$^-$)	Decrease	[8]
Naïve B cells (IgD$^+$CD27$^-$)	Increase	[20]
Double Negative B cells (IgG$^+$/IgA$^+$IgD$^-$CD27$^-$)	Decrease	[20]
Serum IgM	Increase	[20]

in immunity. It has lost its ability to bind peptides due to a definitive closure of the antigen binding cleft that prevents peptide binding and presentation. The HFE protein, expressed on crypt enterocytes of the duodenum, regulates the iron uptake by intestinal cells, having acquired the ability to form complex with the receptor for iron-binding transferring. Mutations in HFE gene are associated with hereditary hemochromatosis, a disorder caused by excessive iron uptake [34,35]. Three common mutations, C282Y, H63D and S65C, have been identified in HFE gene. In particular, the C282Y mutation (a cysteine-to-tyrosine mutation at amino acid 282) destroys its ability to make up a heterodimer with β2-microglobulin. The defective HFE protein fails to associate to the transferring receptor, and the complex cannot be transported to the surface of the duodenal crypt cells. As a consequence, in homozygous people, two to three times the normal amount of iron is absorbed from food by the intestine, resulting in end-organ damage and reducing lifespan. Two other mutations, H63D (a histidine to aspartate at amino acid 63) and S65C (a serine to cysteine at amino acid 65), are associated with milder forms of this disease [34,35].

An association between C282Y mutation and longevity characterizes the Sicilian population studied [33]. In particular, women carriers of C282Y mutation had a higher frequency among the oldest old compared to control women (Table 3). Thus, the C282Y mutation may confer a selective advantage in terms of longevity in Sicilian women. Considering the historical and social context in which the generation of women under study lived, our data seem to propose that the possession of iron-sparing alleles significantly increases the possibility for women to reach longevity. For instance, in Sicily, many pregnancies and an iron-poor diet, consisting mainly in grains, vegetables, and fruits, were still the rule for women born at the beginning of last century. In fact, meat was available for men but not for women; this clearly explains how genetic background also interacts with culture habits [30,31,33].

Our data, showing the relevance of C282Y for women survival to late age, allow adding another piece of evidence to the complex puzzle of genetic and environmental factors involved in control of lifespan in humans. The complex interaction of environmental, historical and genetic factors, differently characterizing the various parts of a country, i.e. Italy, likely plays an important role in determining the gender-specific probability of attaining longevity [30,31,33,36].

Role of innate immunity genes in longevity: the paradigmatic case of TLR4, CCR5, COX-2 and 5-LO genes

According to evolutionary ageing theories, most of the parameters influencing immunosenescence appear to be under genetic control [32,37,38]. An example is given by the innate immune system, involved in neutralizing infectious agents [39]. It plays a beneficial role until the time of reproduction and parental care. In old age, a period largely not foreseen by evolution, it can determine an opposite and detrimental effect through chronic inflammatory responses ("antagonistic pleiotropy") [38,40]. Genetic pro-/anti-inflammatory variations in innate immune response are, indeed, thought to influence the susceptibility of age-related human diseases, by altering host response to environmental and endogenous stress [41]. Thus, they are able to determine a negative or positive control of inflammation, by affecting both interactions between host and microbes and survival of the individual and attainment of longevity. Furthermore, they appear both to be responsible, at least in large part, for different men and women strategies to achieve longevity, and to contribute to the preferential sex dimorphism of the age-related diseases [30,31,33].

From our investigations in Sicilian population, TLR4, CCR5, Cox2, 5-Lo genes can be considered good examples. They provide an ideal model to understand the different implications of their genetic variants in the risk of age-related diseases, i.e. atherosclerosis and prostate cancer (PC), and reciprocally in increased chance to attain longevity.

TLR4 gene (number accession of GenBank: NM-138554.1) codifies the best understood TLR member involved in recognition of LPS, the prototypic TLR4 ligand, and other exogenous and endogenous (i.e. HSPs, hyaluronic acid, β-defensin-2, ox-LDL, fibronectin and amyloid peptide) ligands. TLR4 activation implies a downstream signaling mediated by several intracellular adaptor molecules and the consequent activation of transcription

29

Table 3 Data from our investigations in Sicilian population

Gene	Alleles of genetic variants	Centenarians	Young controls (< 55 years)		P
		N = 35 females	N = 106 females		
HFE	C282	47 (84%)	132 (0%)		8.3×10^{-5} [33]
	282Y	9 (16%)	0 (0%)		

Genes	Alleles of SNPs or genetic variants	Centenarians	Young controls (< 55 years)	MI patients (< 55 years)	P
		N = 55 males	N = 127 males	N = 105 males	
TLR4	+896A	94 (85.4%)	239 (94.1%)	205 (97.6%)	< 0.001 [49]
	+896 G	16 (14.6%)	15 (5.9%)	5 (2.4%)	
		N = 123 males	N = 136 males	N = 133 males	
CCR5	WT	221(89.8%)	252 (92.6%)	263 (98.8%)	0.00006 [48]
	Δ32	25 (10.2%)	20 (7.4%)	3 (1.2%)	
		N = 96 males	N = 170 males	N = 140 males	
Cox-2	-765 G	122 (63.5%)	240 (70.6%)	232(82.8%)	0.000007 [47]
	-765 C	70(36.5%)	100(29.4%)	48(17.2%)	
5-Lo	-1708 G	180 (93.7%)	302(88.8%)	224(80%)	0.00003 [47]
	-1708A	12(6.3%)	38(11.2%)	56(20%)	0.001
	21 C	176(91.7%)	299(88%)	225(80.4%)	
	21 T	16(8.3%)	41(12%)	55(19.6%)	

Genes	Alleles of SNPs or genetic variants	Centenarians	Young controls (< 55 years)	PC patients (< 55 years)	P
		N = 55 males	N = 125 males	N = 50 males	
TLR4	+896A	94 (85%)	235 (94%)	99 (99%)	0.001 [54]
	+896 G	16 (15%)	15 (6%)	1 (1%)	
Cox-2	-765 G	67 (61%)	176 (70%)	77 (77%)	0.05
5-Lo	-765 C	43 (39%)	74 (30%)	23(23%)	0.0007
	-1708 G	104 (95%)	223 (89%)	77 (77%)	
	-1708A	6 (5%)	27 (11%)	23 (23%)	
		N = 53 males		N = 50 males	
CCR5	WT	95 (89.6%)		97 (97%)	0.03 [53]
	Δ32	11(10.4%)		3(3%)	

Gene	Alleles of SNPs or genetic variants	Centenarians	Young controls (< 55 years)	BF patients (30-60 years)	P
		N = 42 females	N = 42 females	N = 42 females	
TLR4	+896A	81 (96.4%)	78 (92.9%)	76 (90.4%)	0.003 [55]
	+ 896 G	3 (3.6%)	6 (7.1%)	8 (9.6%)	

factors, such as NF-kB. This determines the production of different pro/anti-inflammatory mediators. These lasts, such as IL-10, are produced by the parallel activation of anti-inflammatory pathways to limit the potential tissue damage from excessive activation of the innate immune system [42]. SNPs seem to modulate both TLR4 activity and function. In human, only two SNPs, +896A/G (Asp299Gly; rs4986790) and +1196 C/T (Thr399Ile; rs4986791), have a frequency > 5%. They induce a blunted response to LPS, as first suggested by Arbour et al., and are phenotypically associated to changes in the production of cytokines, principally those carrying the Asp299Gly mutation [43-45]. Accordingly, recent literature data suggest the ability of this SNP to modulate the risk of major age-related diseases [42].

The CCR5 gene (number accession of GenBank: NM-00579) codifies for a G protein-coupled chemokine receptor, which regulates trafficking and effector functions of memory/effector Th1 cells, macrophages, NK cells and immature dendritic cells. CCR5 and its ligands are important molecules in viral pathogenesis. Recent evidence has also demonstrated the role of CCR5 in a variety of human diseases, ranging from infectious and inflammatory age-related diseases to cancer. A notable variant of CCR5 gene is a 32 bp (Δ32) deletion, which causes a frame shift mutation in exon 4 (CCR5Δ32; rs333) and determines stop protein maturation and loss of expression of functional CCR5 receptor [46]. Accordingly, it seems to have a protective role against CVD and other age-related diseases, such as PC. It, indeed, determines a slower progression of atherogenesis or cancerogenesis as a consequence of an attenuated inflammatory response.

COX-2 gene maps in the 1q25 chromosome and codifies for the Cox-2 enzyme involved in the conversion of arachidonic acid to prostaglandins. Polymorphisms regulate its expression and hence prostanoid biosynthesis. In particular, it has been identified a guanine to cytosine substitution at position -765 G/C, located within a putative binding site for the transcription factor Sp1, associated to a reduction in the risk of clinical cardiovascular events. COX-2 is expressed at low levels in most tissues, but its expression enhances under inflammatory stimuli and in inflammatory age-related processes, i.e. atherosclerosis, rheumatoid diseases and cancer [47].

The 5-LO gene maps in the chromosome 10q11.2 and codifies the 5-Lo enzyme involved in the synthesis of LTs. The 5-LO pathway has been associated to atherosclerosis in mouse and human histological studies. Several SNPs have been described. In particular, the - 1708 G/A, -1761 G/A and 21 C/T SNPs in promoter region and exon-1 of 5-LO gene modify the gene transcription or the putative protein [47].

An over-expression of anti-inflammatory CCR5Δ32 variant, +896 G (299Gly) TLR4 allele, -765 C Cox-2 allele, -1708 G and 21 C 5-Lo alleles characterizes male Sicilian centenarians (Table 3) [47-49]. So, male centenarians are people who seem genetically equipped for defeating major age-related diseases. They present SNPs in the immune system genome (i.e. SNPs or other genetic variations, located within the promoter regions of pro-inflammatory cytokines) which, regulating the immune-inflammatory responses, seem to be associated to longevity [30-32]. Furthermore, centenarians are characterized by marked delay or escape from age-associated diseases, responsible for the high mortality in earlier ages. In particular, it has been demonstrated that centenarian offspring have an increased likelihood of surviving to 100 years and show a reduced prevalence of age associated diseases, such as CVD, and lower prevalence of CVD risk factors [1,30-32,50] Thus, genes involved in CVD may play an opposite role in human male longevity. Our data in male Sicilian population confirm this suggestion and emphasize the role of antagonistic pleiotropy in ageing and longevity [51,52]. A high frequency of proinflammatory CCR5wt variant, +896A TLR4 allele, -765 G Cox-2 allele, 1708A and 21 T 5-Lo alleles characterizes male Sicilian patients affected by MI (Table 3) [47-49]. In a recent study, we also found a similar overexpression of these proinflammatory SNPs in male Sicilian patients affected by PC (Table 3). Opposite data were obtained in male centenarians [53,54].

In contrast, female Sicilian centenarians have a different frequency of the alleles of +896A/G TLR4 SNP than that observed in male Sicilian centenarians. In particular, female Sicilian centenarians show an over-expression of the pro-inflammatory +896A TLR4 allele respect to female patients affected by Boutonneuse fever and age-matched controls (Table 3) [55].

On the other hand, pro-inflammatory responses are evolutionary programmed to resist fatal infections. Thus, it is not surprising that the genetic background of people that survive to an advanced age may be protective against infections [55].

Based on our data, we suggest that Sicilian men and women may follow different trajectories to reach longevity. For men it might be more important to control atherogenesis and cancerogenesis, whereas for women it might be more important to control infectious diseases [30,31].

In order to confirm our suggestions on the biological effects of +896A/G TLR4 SNP and its role in the patho-physiology of age-related diseases studied (i.e. MI and PC) and longevity, we recently assessed the levels of IL-6, TNF-α, IL-10 and eicosanoids in LPS-stimulated whole blood samples from 50 young healthy Sicilians, screened for the presence of this SNP. Both pro-inflammatory cytokines and eicosanoids were significantly lower in carriers bearing the +896 G TLR4 allele, whereas the anti-inflammatory IL-10 values were higher [56]. This suggests the ability of the +896 G TLR4 allele to mediate a better control of inflammatory responses induced by chronic stimuli, so likely decreasing the effects of athero-genetic damage and prostate carcinogens.

On the basis of data reported herein, some suggestions can be drawn. First, pathogen load, by interacting with the host genotype, determines the type and intensity of inflammatory responses, according to the pro-inflammatory status and tissue injury, implicated in the patho-physiology of major age-related diseases. Second, adequate control of inflammatory response might reduce the risk of these diseases, and, reciprocally, might increase the chance of extended survival in an environment with reduced pathogen load. Accordingly, a higher frequency of the anti-inflammatory +896 G TLR4 allele has been observed in centenarians [49].

Cytokine profile: a biomarker for successful ageing

Cytokines are considered key players in maintaining lymphocyte homeostasis [57,58]. Their function is not limited to induce response after an immune insult, but they can modulate the nature of response (cytotoxic, humoral, cell mediated, inflammatory or allergic) or, in contrast, they may cause non-responsiveness and active immune suppression [58]. Furthermore, sequence variations in several cytokine genes, such as IFN-γ and IL-10 genes, have been demonstrated to be associated with successful ageing and longevity [58]. On the other hand, individual changes in type and intensity of immune response affecting life span expectancy and health ageing seem to have a genetic component. A well-preserved immune function characterizing the successful ageing has been found in centenarians [38]. Recent evidence suggests that centenarians seem to be genetically equipped gene polymorphism for overcame the major age-related diseases and polymorphisms in immune system genes involved in regulation of immune responses have been found associated to longevity. In particular, associations between both cytokine gene polymorphisms and longevity, and differential gender longevity in males and females, and reciprocally to age-related diseases have been demonstrated [38,58,59].

Our data in Sicilian population confirm these associations and suggest that differences in the genetic regulation of immune inflammatory processes might explain the reason why some people but not others develop age-related diseases and why some develop a greater inflammatory response than others. In particular, this suggestion seems to be suitable for some SNPs in IFN-γ and IL-10 genes (Table 4) [60-63].

IFN-γ gene codifies for a cytokine involved in defense against viruses and intracellular pathogens, and in induction of immune mediated inflammatory responses. Its production is genetically regulated. A variable length CA repeat sequence in the first intron of IFN-γ gene has been described to be associated with high IFN-γ production. Furthermore, a SNP, T to A (+874 T/A), at 59 end of the CA repeat region has been described and T presence has been related to high-producing microsatellite allele 2. This SNP coincides with a putative NF-κB

binding site, which might have functional consequences for transcription of IFN-γ gene. Thus, this SNP might directly influence IFN-γ production levels associated to CA microsatellite marker [60].

IL-10 gene codifies for IL-10 cytokine. IL-10 is produced by macrophages, T and B cells. It is one of the major immune-regulatory cytokines, usually considered to mediate potent down-regulation of inflammatory responses. IL-10 production, independently on interaction with other cytokine gene products, is generally controlled by several polymorphic elements in the 5' flanking region of IL-10 gene. Multiple SNPs have been identified in human IL-10 5' flanking region and some of these (i.e. -592, -819, -1082) combine with microsatellite alleles to form haplotype associated with differential IL-10 production. These three SNPs in the IL-10 proximal gene region (considered potential targets for transcription regulating factors) might be involved in genetic control of IL-10 production, even if contrasting literature data have been reported. In particular, the homozygous -1082GG genotype seems to be associated with higher IL-10 production respect to G/A heterozygous and AA homozygous genotypes. Furthermore, this SNP seems to be functionally relevant. It has been demonstrated that -1082 A carriers (low producers) seem likely develop a major number of chronic inflammatory diseases [61-63].

Our results demonstrated an increase of subjects carrying the -1082 G IL-10 allele in centenarian men [61-63]. This allele is associated to significantly increased IL-10 production. Conversely, we observed that the frequency of -1082A allele, associated to low IL-10 production, was significantly higher in MI patients (Table 4) [63]. Thus,

Table 4 Cytokine data from our studies in Sicilian population

Gene	Genotypes	Centenarians	Young controls (< 55 years)		P
		N = 31 males	N = 161 males		
IL-10	-1082GG	18 (58%)	55 (34%)		< 0.025 [61]
	-1083GA	9 (29%)	88 (55%)		
	-1082AA	4 (13%)	18 (11%)		
		N = 72 males	N = 115 males		
IL-10	-1082GG	33 (46%)	32(28%)		0.019 [62]
	-1083GA	34(47%)	64(56%)		
	-1082AA	5(7%)	19(16%)		
		Centenarians	Young controls (< 55 years)	MI patients (< 55 years)	P
		N = 52 males	N = 110 males	N = 90 males	
IL-10	-1082GG	25 (48.1%)	26(23.6%)	17 (18.9%)	0.003 [63]
	-1083GA	23 (44.2%)	56 (50.9%)	29 (32.2%)	
	-1082AA	6(11.5%)	28 (25.5%)	44 (48.9%)	
Genes	Alleles of SNP	Centenarians	Young controls (< 55 years)		P
		N = 142 females	N = 90 females		
IFN-γ	+874 T	102 (35.9%)	85 (47.2%)		0.02 [60]
	+ 874A	182 (64.1%)	95 (52.8%)		

high IL-10 production seems to be protective vs. MI and a possible biomarker for longevity. People with exceptional longevity have genetic factors (i.e. protective factors for CVD) that modulate ageing processes [63]. This supports the opinion that a genetic background protective against CVD is a component of longevity. On the other hand, our immune system has evolved to control pathogens and pro-inflammatory responses are likely programmed by evolution to resist fatal infections. From this prospective, low IL-10 production is correlated with increased resistance to pathogens. In older ages not evolutionally programmed, increased IL-10 levels might better control inflammatory responses induced by chronic vessel damage and reduce the risk for atherogenetic complications. These conditions might permit to achieve exceptional ages in an environmental with a reduced pathogen load [63].

In contrast, female Sicilian centenarians are characterized by an over-expression of +874 INF-γ allele (Table 4) [60]. The INF-γ production is also influenced by hormonal control fundamentally mediated by 17β extradiol. Hormonal regulation of this cytokine has been suggested to modulate, in part, the ability of estrogens to potentiate many types of immune responses and to influence the disproportionate susceptibility of women for immune-inflammatory diseases. Thus, gene variants representing genetic advantage for one gender might not be reciprocally relevant for the other gender in terms of successful or unsuccessful ageing [60].

The data from Sicilian investigation add another piece to complex puzzle of genetic and environmental factors involved in the control of life span expectancy in humans. Studies on cytokine gene SNPs may promise to individuate a complex network of trans-inactive genes able to influence the type and strength of immune responses to environmental stressors, and as final result, conditioning individual life expectancy [60-63]. On the other hand, we recently suggested the possibility to use cytokine profile as biomarker of successful ageing, by evaluating through Lumines technology cytokine serum levels in 44 Sicilian nonagenarians and 79 control subjects (aged between 30 and 50 years old) [64]. IFN-γ and IL-2 levels are unmodified, suggesting a substantial maintenance of relevant T functions. In addition, a significant increase of IL-12 serum levels was observed. This condition might be associated with the increase of NK cell function with ageing. Furthermore, an increase of IL-13 and a reduction of IL-4 were found. Thus, the maintenance of some effector's mechanisms of immune-response characterizes advanced ages. From a general point of view, our data firstly confirm the age-related remodeling of cytokine network. Furthermore, they underline the presence of unchanged levels of some crucial cytokines useful in preserving key immune function in long-living persons [64].

Future perspectives

The ageing process and longevity are multi-factorial events. Genetic, epigenetic, stochastic and environmental factors seem to have a crucial role in ageing and longevity. Epigenetic is associated to ageing, as shown in the major number of studies. In particular, ageing is associated to a global loss of methylation state [65]. In addition, tissue-dependent age-related hypermetylation of specific DNA regions have been observed. Thus, it can be concluded that epigenetic age-related modification are stochastic and no linked to specific DNA region, while epigenetic changes linked to specific environmental stimuli are limited in specific DNA region [66,67]. These observations have led to address the research on epigenomics and its implication in ageing and longevity.

Epigenomics is the systematic study of the global gene expression changes due to epigenetic processes, but not to DNA base sequence changes. Epigenetic processes consist in heritable modification that result in a selective gene expression or repression and consequently in phenotype changes [68]. These changes include nucleosome positioning, post-translation histone modifications, action of small RNAs, DNA replication timing, heterochromatinization and DNA methylation [69]. This last one consists in the addition of a methyl group (-CH3) in the carbon 5 of cytosines, particularly in the CpG dinucleotide. This condition particularly concerns the CpG islands (CpGIs), located at the regulatory site of gene promoter regions. Methylation rate is associated to transcriptional regulation. In particular, gene silencing is associated to increase of -CH3 groups on DNA, conversely hypometylation of CGIs is associated to an open chromatin state resulting in gene expression [70].

Although the association between ageing and epigenetic is a real evidence, processes involved are not clear. Certainly, the nutrition affects epigenetic modifications. Nutrients can be active on specific sites. For example, vitamin B12, vitamin B6, riboflavin, methionine, choline and betaine, well known as folates, regulate levels of S-adenosylmethionine and S-adenosylhomocysteine, donor of -CH3 group and methyltransferase inhibitor respectively [71]. Curcumin, resveratrol, polyphenols and flavonoids, phytoestrogen, and lycopene are also considered key nutritional factors both for regulation of enzyme involved in acetylation and deacetylation mechanism and for one-carbon metabolism [71,72]. A diet rich in vegetables and fruit, such as Mediterranean diet, may contain these nutrients. Sicilian centenarians are characterized to observe this kind of diet, as we reported [73]. Since genetic and environmental factors contribute to longevity, it may suggest that epigenetic events associated to the modifications diet-induced are very important for successful ageing processes. Furthermore, several literature data reported a possible link between epigenetic and several age-related diseases, such as cancer, metabolic syndrome, diabetes and neurodegenerative disorders. Stable propagation of gene expression from cell to cell during disease pathogenesis is regulated by epigenetic mechanisms. For example, during the diabetes onset epigenetic changes act on insulin and insulin metabolism regulating the gene coding [74]. In particular, a recent study has demonstrated that human insulin gene and mouse insulin 2 gene expression are under control of epigenetic changes in CpGIs. Insulin non expressing cells are, indeed, methylated in the promoter region of insulin coding gene, while insulin expressing cells are completely demethylated in the same site resulting in insulin gene expression [75]. Another study on monozygotic twin has demonstrated that insulin resistance is also under control of DNA methylation [76]. Alterations in insulin pathway are known to be involved in metabolic disease, such as metabolic syndrome, insulin resistance and type 2 diabetes. Recent data also support the existence of a correlation between these alterations and Alzheimer's disease.

In the light of these observations, the purpose of our future studies will be to evaluate the weight of epigenetic changes in ageing and longevity, using centenarians as super-controls.

Abbreviations

AD: Alzheimer's disease; BF: Boutonnese fever; CCR5: CC chemokine receptor 5; COX-2: Cyclo-oxygenase 2; CRP: C reactive protein; CVD: Cardiovascular disease; HSPs: Heat-shock proteins; INF-γ: Interferon- γ; IL-6: Interleukin-6; IL-10: Interleukin-10; 5-LO: 5-lipoxygenase; LPS: Lipopolysaccharide; LTs: Leukotrienes; MI: Myocardial infarction; ox-LDL: Oxidized-Low Density Lipoproteins; PC: Prostate cancer; PGs: Prostaglandins; SNPs: Single nucleotide polymorphisms; TLR4: Toll-like-receptor-4; TNF-α: Tumor necrosis factor-α.

Acknowledgements

GA, AM, LV are PhD students and SB is a PhD candidate of Pathobiology PhD course (directed by Prof Calogero Caruso) at Palermo University and this paper is submitted in partial fulfillment of requirement for their PhD degree.

Author details

[1]Department of Pathobiology and Medical and Forensic Biotechnologies, University of Palermo, Corso Tukory 211, Palermo 90134, Italy. [2]Department of Internal Medicine II, Center for Medical Research, University of Tübingen, Tübingen Ageing and Tumor Immunology group, Waldhörnlestr. 22, Tübingen 72072, Germany.

Authors' contributions

CRB, GA, SB, MB, AM and GCR wrote the first draft. Subsequent drafts were written by CRB, who had the overall supervision of the review processing. All authors edited the paper and approved its final version.

Competing interests

The authors declare that they have no competing interests.

Received: 29 March 2012 Accepted: 23 April 2012
Published: 23 April 2012

References

1. Terry DF, Wilcox MA, McCormick MA, Pennington JY, Schoenhofen EA, Andersen SL, Perls TT: **Lower all-cause, cardiovascular, and cancer mortality in centenarians' offspring.** *J Am Geriatr Soc* 2004, **52**:2074-2076.

2. Terry DF, McCormick M, Andersen S, Pennington J, Schoenhofen E, Palaima E, Bausero M, Ogawa K, Perls TT, Asea A: **Cardiovascular disease delay in centenarian offspring: role of heat shock proteins.** *Ann N Y Acad Sci* 2004, **1019**:502-505.

3. Barzilai N, Gabriely I, Gabriely M, Iankowitz N, Sorkin JD: **Offspring of centenarians have a favorable lipid profile.** *J Am Geriatr Soc* 2001, **49**:76-79.

4. Rose G, Passarino G, Scornaienchi V, Romeo G, Dato S, Bellizzi D, Mari V, Feraco E, Maletta R, Bruni A, Franceschi C, De Benedictis G: **The mitochondrial DNA control region shows genetically correlated levels of heteroplasmy in leukocytes of centenarians and their offspring.** *BMC Genomics* 2007, **8**:293.

5. Pawelec G, Akbar A, Caruso C, Solana R, Grubeck-Loebenstein B, Wikby A: **Human immunosenescence: is it infectious?** *Immunol Rev* 2005, **205**:257-268.

6. Chidrawar S, Khan N, Wei W, McLarnon A, Smith N, Nayak L, Moss P: **Cytomegalovirus-seropositivity has a profound influence on the magnitude of major lymphoid subsets within healthy individuals.** *Clin Exp Immunol* 2009, **155**:423-432.

7. Derhovanessian E, Larbi A, Pawelec G: **Biomarkers of human immunosenescence: impact of Cytomegalovirus infection.** *Curr Opin Immunol* 2009, **21**:440-445.

8. Derhovanessian E, Maier AB, Beck R, Jahn G, Hähnel K, Slagboom PE, de Craen AJ, Westendorp RG, Pawelec G: **Hallmark features of immunosenescence are absent in familial longevity.** *Immunol* 2010, **185**:4618-4624.

9. Colonna Romano G, Bulati M, Aquino A, Scialabba G, Candore G, Lio D, Motta M, Malaguarnera M, Caruso C: **B cells in the aged: CD27, CD5 and CD40 expression.** *Mech Ageing Dev* 2003, **124**:389-393.

10. Colonna Romano G, Cossarizza A, Aquino A, Scialabba G, Bulati M, Lio D, Candore G, Di Lorenzo G, Fradà G, Caruso C: **Age- and gender-related values of lymphocyte subsets in subjects from northern and southern Italy.** *Arch Gerontol Geriatr (Suppl)* 2002, **8**:99-107.

11. Listì F, Candore G, Modica MA, Russo MA, Di Lorenzo G, Esposito-Pellitteri M, Colonna Romano G, Aquino A, Bulati M, Lio D, Franceschi C, Caruso C: **A study of serum immunoglobulin levels in elderly persons that provides new insights into B cell immunosenescence.** *Ann N Y Acad Sci* 2006, **1089**:487-495.

12. Candore G, Di Lorenzo G, Mansueto P, Melluso M, Fradà G, Li Vecchi M, Esposito Pellitteri M, Drago A, Di Salvo A, Caruso C: **Prevalence of organ-specific and non organ-specific autoantibodies in healthy centenarians.** *Mech Ageing Dev* 1997, **94**:183-190.

13. Colonna-Romano G, Bulati M, Aquino A, Vitello S, Lio D, Candore G, Caruso C: **B cell immunosenescence in the elderly and in centenarians.** *Rejuvenation Res* 2008, **11**:433-439.

14. Colonna-Romano G, Bulati M, Aquino A, Pellicanò M, Vitello S, Lio D, Candore G, Caruso C: **A double-negative (IgD-CD27-) B cell population is increased in the peripheral blood of elderly people.** *Mech Ageing Dev* 2009, **130(10)**:681-690.

15. Bulati M, Buffa S, Candore G, Caruso C, Dunn-Walters DK, Pellicanò M, Wu YC, Colonna Romano G: **B cells and immunosenescence: a focus on IgG+IgD-CD27- (DN) B cells in aged humans.** *Ageing Res Rev* 2011, **10**:274-284.

16. Buffa S, Bulati M, Pellicanò M, Dunn-Walters DK, Wu YC, Candore G, Vitello S, Caruso C, Colonna-Romano G: **B cell immunosenescence: different features of naive and memory B cells in elderly.** *Biogerontology* 2011, **12**:473-483.

17. Wei C, Anolik J, Cappione A, Zheng B, Pugh-Bernard A, Brooks J, Lee EH, Milner EC, Sanz I: **A new population of cells lacking expression of CD27 represents a notable component of the B cell memory compartment in systemic lupus erythematosus.** *J Immunol* 2007, **178**:6624-6633.

18. Sanz I, Wei C, Lee FE, Anolik J: **Phenotypic and functional heterogeneity of human memory B cells.** *Semin Immunol* 2008, **20(1)**:67-82.

19. Cagigi A, Du L, Dang LV, Grutzmeier S, Atlas A, Chiodi F, Pan-Hammarström Q, Nilsson A: **CD27(-) B-cells produce class switched and somatically hyper-mutated antibodies during chronic HIV-1 infection.** *PLoS One* 2009, **4(5)**:e5427.

20. Colonna-Romano G, Buffa S, Bulati M, Candore G, Lio D, Pellicanò M, Vasto S, Caruso C: **B cells compartment in centenarian offspring and old people.** *Curr Pharm Des* 2010, **16**:604-608.

21. Frasca D, Riley RL, Blomberg BB: Humoral immune response and B-cell functions including immunoglobulin class switch are downregulated in aged mice and humans. *Semin Immunol* 2005, 17(5):378-384.
22. Dunn-Walters DK, Ademokun AA: B cell repertoire and ageing. *Curr Opin Immunol* 2010, 22(4):514-520.
23. Cancro MP, Hao Y, Scholz JL, Riley RL, Frasca D, Dunn-Walters DK, Blomberg BB: B cells and aging: molecules and mechanisms. *Trends Immunol* 2009, 30(7):313-318.
24. Kumar R, Burns EA: Age-related decline in immunity: implications for vaccine responsiveness. *Expert Rev Vaccines* 2008, 7(4):467-479.
25. Gibson KL, Wu YC, Barnett Y, Duggan O, Vaughan R, Kondeatis E, Nilsson BO, Wikby A, Kipling D, Dunn-Walters DK: B-cell diversity decreases in old age and is correlated with poor health status. *Aging Cell* 2009, 8(1):18-25.
26. Grubeck-Loebenstein B, Della Bella S, Iorio AM, Michel JP, Pawelec G, Solana R: Immunosenescence and vaccine failure in the elderly. *Aging Clin Exp Res* 2009, 21(3):201-209.
27. Perrig-Chiello P, Hutchison S: Health and well-being in old age: the pertinence of a gender mainstreaming approach in research. *Gerontology* 2010, 56:208-213.
28. May RC: Gender, immunity and the regulation of longevity. *Bioessays* 2007, 29(8):795-802.
29. [http://www.istat.it].
30. Candore G, Balistreri CR, Colonna-Romano G, Lio D, Listì F, Vasto S, Caruso C: Gender-related immune-inflammatory factors, age-related diseases, and longevity. *Rejuvenation Res* 2010, 13:292-729.
31. Candore G, Balistreri CR, Listì F, Grimaldi MP, Vasto S, Colonna-Romano G, Franceschi C, Lio D, Caselli G, Caruso C: Immunogenetics, gender, and longevity. *Ann N Y Acad Sci* 2006, 1089:516-537.
32. Capri M, Salvioli S, Monti D, Caruso C, Candore G, Vasto S, Olivieri F, Marchegiani F, Sansoni P, Baggio G, Mari D, Passarino G, De Benedictis G, Franceschi C: Human longevity within an evolutionary perspective: The peculiar paradigm of a post-reproductive genetics. *Exp Gerontol* 2008, 43:53-60.
33. Lio D, Balistreri CR, Colonna-Romano G, Motta M, Franceschi C, Malaguarnera M, Candore G, Caruso C: Association between the MHC class I gene HFE polymorphisms and longevity: a study in Sicilian population. *Genes Immun* 2002, 3:20-24.
34. Beutler E: Iron storage disease: facts, fiction and progress. *Blood Cells Mol Dis* 2007, 39:140-147.
35. Pietrangelo A: Hereditary hemochromatosis: pathogenesis, diagnosis, and treatment. *Gastroenterology* 2010, 139:393-408, 408.e1-2.
36. Franceschi C, Motta L, Motta M, Malaguarnera M, Capri M, Vasto S, Candore G, Caruso C: IMUSCE. The extreme longevity: the state of the art in Italy. *Exp Gerontol* 2008, 43:45-52.
37. Troen BR: The biology of aging. *Mt Sinai J Med* 2003, 70:3-22.
38. Ostan R, Bucci L, Capri M, Salvioli S, Scurti M, Pini E, Monti D, Franceschi C: Immunosenescence and immunogenetics of human longevity. *Neuroimmunomodulation* 2008, 15:224-240.
39. Candore G, Colonna-Romano G, Balistreri CR, Di Carlo D, Grimaldi MP, Listì F, Nuzzo D, Vasto S, Lio D, Caruso C: Biology of longevity: role of the innate immune system. *Rejuvenation Res* 2006, 9:143-148.
40. Williams GC: Pleiotropy, natural selection, and the evolution of senescence. *Evolution* 1957, 398-411, 11.
41. Vasto S, Candore G, Balistreri CR, Caruso M, Colonna-Romano G, Grimaldi MP, Listi F, Nuzzo D, Lio D, Caruso C: Inflammatory networks in ageing, age-related diseases and longevity. *Mech Ageing Dev* 2007, 128:83-91.
42. Balistreri CR, Colonna-Romano G, Lio D, Candore G, Caruso C: TLR4 polymorphisms and ageing: implications for the pathophysiology of age-related diseases. *J Clin Immunol* 2009, 29:406-415.
43. Arbour NC, Lorenz E, Schutte BC, Zabner J, Kline JN, Jones M, Frees K, Watt JL, Schwartz DA: TLR4 mutations are associated with endotoxin hyporesponsiveness in humans. *Nat Genet* 2000, 25:187-191.
44. Ferwerda B, McCall MB, Alonso S, Giamarellos-Bourboulis EJ, Mouktaroudi M, ùlzagirre N, Syafruddin D, Kibiki G, Cristea T, Hijmans A, Hamann L, Israel S, ElGhazali G, Troye-Blomberg M, Kumpf O, Maiga B, Dolo A, Doumbo O, Hermsen CC, Stalenhoef AF, van Crevel R, Brunner HG, Oh DY, Schumann RR, de la Rúa C, Sauerwein R, Kullberg BJ, van der Ven AJ, van der Meer JW, Netea MG: TLR polymorphisms, infectious diseases, and
45. Ferwerda B, McCall MB, Verheijen K, Kullberg BJ, van der Ven AJ, Van der Meer JW, Netea MG: Functional consequences of toll-like receptor 4 polymorphisms. *Mol Med* 2008, 14:346-352.
46. Balistreri CR, Caruso C, Grimaldi MP, Listì F, Vasto S, Orlando V, Campagna AM, Lio D, Candore G: CCR5 receptor: biologic and genetic implications in age-related diseases. *Ann NY Acad Sci* 2007, 1100:162-172.
47. Listì F, Caruso M, Incalcaterra E, Hoffmann E, Caimi G, Balistreri CR, Vasto S, Scafidi V, Caruso C, Candore G: Pro-inflammatory gene variants in myocardial infarction and longevity: implications for pharmacogenomics. *Curr Pharm Des* 2008, 14:2678-2685.
48. Balistreri CR, Candore G, Caruso M, Incalcaterra E, Franceschi C, Caruso C: Role of polymorphisms of CC-chemokine receptor-5 gene in acute myocardial infarction and biological implications for longevity. *Haematologica* 2008, 93:637-638.
49. Balistreri CR, Candore G, Colonna-Romano G, Lio D, Caruso M, Hoffmann E, Franceschi C, Caruso C: Role of Toll-like receptor 4 in acute myocardial infarction and longevity. *JAMA* 2004, 292:2339-2340.
50. Perls T, Terry D: Genetics of exceptional longevity. *Exp Gerontol* 2003, 38:725-730.
51. Candore G, Balistreri CR, Grimaldi MP, Listì F, Vasto S, Caruso M, Caimi G, Hoffmann E, Colonna-Romano G, Lio D, Paolisso G, Franceschi C, Caruso C: Opposite role of pro-inflammatory alleles in acute myocardial infarction and longevity: results of studies performed in a Sicilian population. *Ann NY Acad Sci* 2006, 1067:270-275.
52. Incalcaterra E, Caruso M, Balistreri CR, Candore G, Lo Presti R, Hoffmann E, Caimi G: Role of genetic polymorphisms in myocardial infarction at young age. *Clin Hemorheol Microcirc* 2010, 46:291-298.
53. Balistreri CR, Carruba G, Calabrò M, Campisi I, Di Carlo D, Lio D, Colonna-Romano G, Candore G, Caruso C: CCR5 proinflammatory allele in prostate cancer risk: a pilot study in patients and centenarians from Sicily. *Ann NY Acad Sci* 2009, 1155:289-292.
54. Balistreri CR, Caruso C, Carruba G, Miceli V, Campisi I, Listì F, Lio D, Colonna-Romano G, Candore G: A pilot study on prostate cancer risk and pro-inflammatory genotypes: pathophysiology and therapeutic implications. *Curr Pharm Des* 2010, 16:718-724.
55. Balistreri CR, Candore G, Lio D, Colonna-Romano G, Di Lorenzo G, Mansueto P, Rini G, Mansueto S, Cillari E, Franceschi C, Caruso C: Role of TLR4 receptor polymorphisms in Boutonneuse fever. *Int J Immunopathol Pharmacol* 2005, 18:655-660.
56. Balistreri CR, Caruso C, Listì F, Colonna-Romano G, Lio D, Candore G: LPS-mediated production of pro/anti-inflammatory cytokines and eicosanoids in whole blood samples: biological effects of +896A/G TLR4 polymorphism in a Sicilian population of healthy subjects. *Mech Ageing Dev* 2011, 132:86-92.
57. Sanjabi S, Zenewicz LA, Kamanaka M, Flavell RA: Anti-inflammatory and pro-inflammatory roles of TGF-beta, IL-10, and IL-22 in immunity and autoimmunity. *Curr Opin Pharmacol* 2009, 9:447-453.
58. Iannitti T, Palmieri B: Inflammation and genetics: an insight in the centenarian model. *Hum Biol* 2011, 83:531-559.
59. Caruso C, Candore G, Colonna-Romano G, Lio D, Franceschi C: *Inflammation and life-span. Science* 2005, 307:208-209.
60. Lio D, Scola L, Crivello A, Colonna-Romano G, Candore G, Bonafè M, Cavallone L, Franceschi C, Caruso C: Gender-specific association between -1082 IL-10 promoter polymorphism and longevity. *Genes Immun* 2002, 3:30-33.
61. Lio D, Scola L, Crivello A, Colonna-Romano G, Candore G, Bonafé M, Cavallone L, Marchegiani F, Olivieri F, Franceschi C, Caruso C: Inflammation, genetics, and longevity: further studies on the protective effects in men of IL-10-1082 promoter SNP and its interaction with TNF-alpha -308 promoter SNP. *J Med Genet* 2003, 40:296-299.
62. Lio D, Candore G, Crivello A, Scola L, Colonna-Romano G, Cavallone L, Hoffmann E, Caruso M, Licastro F, Caldarera CM, Branzi A, Franceschi C, Caruso C: Opposite effects of interleukin 10 common gene polymorphisms in cardiovascular diseases and in successful ageing: genetic background of male centenarians is protective against coronary heart disease. *J Med Genet* 2004, 41:790-794.
63. Lio D, Scola L, Crivello A, Bonafè M, Franceschi C, Olivieri F, Colonna-Romano G, Candore G, Caruso C: Allele frequencies of +874 T- > A single

nucleotide polymorphism at the first intron of interferon-gamma gene in a group of Italian centenarians. *Exp Gerontol* 2002, **37**:315-319.

64. Palmeri M, Misiano G, Malaguarnera M, Forte GI, Vaccarino L, Milano S, Scola L, Caruso C, Motta M, Maugeri D, Lio D: **Cytokine serum profile in a group of sicilian nonagenarians.** *J Immunoassay Immunochem* 2012, **33**:82-90.

65. Park LK, Friso S, Choi SW: **Nutritional influences on epigenetics and age-related disease.** *Proc Nutr Soc* 2012, **71**:75-83.

66. Portela A, Esteller M: **Epigenetic modifications and human disease.** *Nat Biotechnol* 2010, **28**:1057-1068.

67. Fuke C, Shimabukuro M, Petronis A, Sugimoto J, Oda T, Miura K, Miyazaki T, Ogura C, Okazaki Y, Jinno Y: **Age related changes in 5-methylcytosine content in human peripheral leukocytes and placentas: an HPLC-based study.** *Ann Hum Genet* 2005, **69**(Pt 1):134.

68. Wu C, Morris JR: **Genes, genetics, and epigenetics: a correspondence.** *Science* 2001, **293**:1103-1105.

69. Antequera F, Bird A: *CpG islands. EXS* 1993, **64**:169-185.

70. Decottignies A, d'Adda di Fagagna F: **Epigenetic alterations associated with cellular senescence: a barrier against tumorigenesis or a red carpet for cancer?** *Semin Cancer Biol* 2011, **21**:360-366.

71. Lim U, Song MA: **Dietary and lifestyle factors of DNA methylation.** *Methods Mol Biol* 2012, **863**:359-376.

72. Choi SW, Friso S: **Epigenetics: A New Bridge between Nutrition and Health.** *Adv Nutr* 2010, **1**:8-16.

73. Vasto S, Scapagnini G, Rizzo C, Monastero R, Marchese A, Caruso C: **Mediterranean diet and longevity in Sicily: a survey in Sicani Mountains Population.** *Rejuvenation Res* .

74. Mitić T, Emanueli C: **Diabetes-induced epigenetic signature in vascular cells.** *Endocr Metab Immune Disord Drug Targets* 2012.

75. Kuroda A, Rauch TA, Todorov I, Ku HT, Al-Abdullah IH, Kandeel F, Mullen Y, Pfeifer GP, Ferreri K: **Insulin gene expression is regulated by DNA methylation.** *PLoS One* 2009, **4**:10.

76. Zhao J, Goldberg J, Bremner JD, Vaccarino V: **Global DNA methylation is associated with insulin resistance: a monozygotic twin study.** *Diabetes* 2012, **61**:542-546.

doi:10.1186/1742-4933-9-8
Cite this article as: Balistreri *et al.*: **Genetics of longevity. Data from the studies on Sicilian centenarians.** *Immunity & Ageing* 2012 **9**:8.

Vasto et al. Immunity & Ageing 2012, 9:10
http://www.immunityageing.com/content/9/1/10

IMMUNITY & AGEING

SHORT REPORT   **Open Access**

Centenarians and diet: what they eat in the Western part of Sicily

Sonya Vasto[1,3*], Claudia Rizzo[2,3] and Calogero Caruso[2,3]

Abstract

This paper pays attention to the modifiable lifestyle factors such as diet and nutrition that might influence life extension and successful ageing. Previous data reported that in Sicily, the biggest Mediterranean island, there are some places where there is a high frequency of male centenarians with respect to the Italian average. The present data show that in Sicani Mountain zone there are more centenarians with respect to the Italian average. In fact, in five villages of Sicani Mountains, there were 19 people with an age range of 100–107 years old from a total population of 18,328 inhabitants. So, the centenarian number was 4.32-fold higher than the national average (10.37 vs. 2.4/10,000); the female/male ratio was 1.1:1 in the study area, while the national ratio is 4.54:1. Unequivocally, their nutritional assessment showed a high adherence to the Mediterranean nutritional profile with low glycemic index food consumed. To reach successful ageing it is advisable to follow a diet with low quantity of saturated fat and high amount of fruits and vegetables rich in phytochemicals.

Keywords: Ageing, Centenarian, Longevity, Mediterranean diet

Introduction

The Mediterranean diet has been widely recommended for a healthy lifestyle since Ancel Keys first used the term in 1975 [1]. The essential concept is that this is not a set of changes to our usual diet dictated by scientific experiments, but a set of food habits and recipes traditionally enjoyed by the ordinary people of Mediterranean countries, who have been found to have lower rates of coronary and other age-related chronic diseases, including cancer, than most developed countries [2,3].

In recent years, researchers have been extremely interested in the clear advantage of the Mediterranean nutritional recommendation that exists and is used in the Mediterranean surrounding area. In fact, there is no single Mediterranean diet but several interpretations based on the Mediterranean country's political, economic and cultural tradition [4].

Since 1990, increasing evidence suggests that these diets have a beneficial influence on several diseases such as cardiovascular diseases, metabolic syndromes, hence showing protective effect on health and longevity [5-7]. Mediterranean diet is characterized by a high intake of monounsaturated fat, plant proteins, whole grains (fish is not always present), moderate intake of alcohol, and low consumption of red meat, refined grains, and sweets. Further, the consumption of large amount of olive oil and olives in meals dominates all the Mediterranean cuisine [8].

Historically, the beneficial properties of virgin olive oil were attributed to the high proportion of monounsaturated fatty acids (MUFAs), namely oleic acid, rather than to the phenolic fraction. Nevertheless, several seed oils, including sunflower, soybean, and rapeseed, rich in MUFA have been demonstrated to be ineffective in beneficially altering chronic disease risk factors. Therefore, it is likely that the polyphenols in olive oil may mediate these health benefits [8,9].

There are at least thirty-six structurally distinct phenolics that have been identified in virgin olive oil, but not all phenolic compounds and their concentration are present in every virgin olive oil. Such differences in the phenolic compound are dependent on several factors like the variety of the olive fruit, the region in which the olive fruit is grown, the agricultural techniques used, the maturity of the olive fruit at harvest, the extraction process and the storage method [10].

---

* Correspondence: sonya.vasto@unipa.it
[1]Department of Molecular and Biomolecular Sciences (STEMBIO), University of Palermo, Via Archifari 32, 90213, Palermo, Italy
[3]Immunohaemathology Unit, University Hospital, University of Palermo, Via del Vespro, 127, 90137, Palermo, Italy
Full list of author information is available at the end of the article

The Sicanian Mountains (or Sicani), bordered by Ficuzza wood in the North, Caltanissetta in the East, Salemi in the West and Agrigento to the South represent a very peculiar area where there is a high frequency of centenarians with respect to the Italian average [11,12]. The goal of this study was to characterize the dietary

Mini-Nutritional Assessment MNA™

Last Name:__________ First Name:__________ M.I.___ Sex:____ Date:_____

Age:_____ Weight (kg):_________ Height (cm):_________ Knee Height (cm):_________

Complete the form by writing the numbers in the boxes. Add the numbers in the boxes and compare the total assessment to the Malnutrition Indicator Score.

Anthropometric Assessment

	Points
1. Body Mass Index (BMI) (weight in kg)/(height in m)² a. BMI <19 = 0 points b. BMI to <21 = 1 point c. BMI 21 to <23 = 2 points d. BMI ≥ 23 = 3 points	☐
2. Mid-arm circumference (MAC) in cm a. MAC < 21 = 0.0 points b. MAC 21 ≤ 22 = 0.5 points c. MAC > 22 = 1.0 points	☐.☐
3. Calf circumference (CC) in cm a. CC < 31 = 0 points b. CC ≥ 31 = 1 point	☐
4. Weight loss during last 3 months a. weight loss greater than 3 kg (6.6 lbs) = 0 points b. does not know = 1 point c. weight loss between 1 & 3 kg = 2 points d. no weight loss = 3 points	☐

General Assessment

5. Lives independently (not in a nursing home or hospital a. no = 0 points b. yes = 1 point	☐
6. Takes more than 3 prescription drugs per day a. yes = 0 points b. no = 1 point	☐
7. Has suffered psychological stress or acute disease in the past 3 months a. yes = 0 points b. no = 1 point	☐
8. Mobility a. bed or chair bound = 0 points b. able to get out of bed/chair but does not go out = 1 point c. goes out = 2 points	☐
9. Neuropsychological problems a. severe dementia or depression = 0 points b. mild dementia = 1 point c. no psychological problems = 2 points	☐
10. Pressure sores or skin ulcers a. yes = 0 points b. no = 1 point	☐

Dietary Assessment

	Point
11. How many full meals does the patient eat daily? a. 1 meal = 0 points b. 2 meals = 1 point c. 3 meals = 2 points	☐
12. Selected consumption markers for protein intake • At least one serving of dairy products (milk, cheese, yogurt) per day? yes ☐ no ☐ • Two or more servings of legumes or eggs per week? yes ☐ no ☐ • Meat, fish, or poultry every day? yes ☐ no ☐ a. if 0 or 1 yes = 0.0 points b. if 2 yes = 0.5 points c. if 3 yes = 1.0 points	☐.☐
13. Consumes two or more servings of fruits or vegetables per day? a. no = 0 points b. yes = 1 point	☐
14. Has food intake declined over the past three months due to loss of appetite, digestive problems, chewing or swallowing difficulties? a. severe loss of appetite = 0 points b. moderate loss of appetite = 1 point c. no loss of appetite = 2 points	☐
15. How much fluid (water, juice, coffee, tea, milk...) is consumed per day? (1 cup = 8 ox.) a. less than 3 cups? = 0.0 points b. 3 to 5 cups = 0.5 points c. more than 5 cups = 1.0 points	☐.☐
16. Mode of feeding a. unable to eat without assistance = 0 points b. self-fed with some difficulty = 1 point c. self-fed without any problem = 2 points	☐

Self-Assessment

17. Do they view themselves as having nutritional problems? a. major malnutrition = 0 points b. does not know or moderate malnutrition = 1 point c. no nutritional problem = 2 points	☐
18. In comparison with other people of the same age, how do they consider their health status? a. not as good = 0.0 points b. does not know = 0.5 points c. as good = 1.0 points d. better = 2.0 points	☐.☐

Assessment Total (max. 30 points) ☐☐.☐

Malnutrition Indicator Score

☐ ≥ 24 points well-nourished ☐ 17 to 23.5 points at risk of malnutrition ☐ < 17 points malnourished

Figure 1 Example of MNA administered to healthy centenarians.

habits of centenarians residing around the Sicani mountains, in 5 villages, namely Giuliana, Bisacquino, Castronovo, Chiusa Scalafani, and Prizzi.

Materials and Methods

19 centenarians (10 females and 9 males) living at home in the five municipalities of Bisacquino, Castronovo, Chiusa Scalafani, Giuliana and Prizzi in the Western part of Sicily, Italy, were identified for the present study. These villages are located above sea-level, on the South-Western edge of the Sicani Mountains. Subjects were individuated by general practitioners and their age checked in the birth registries. As a further control, in the interview, particular attention was paid to the concordance between reported age and personal chronologies (age of marriage and of military service for men, age of first and last pregnancy for women, age of children, among others). The subjects underwent a physical examination and a morning fasting blood venous sample was obtained for studying blood chemistry parameters. Anthropometric measures included height, weight, and the body mass index (BMI) [weight (kg)/height (m2)]. Furthermore, the Mini Nutritional Assessment (MNA), Basic Activities of Daily Living (ADL) and the Instrumental Activities of Daily Living (IADL) were administrated. ADL and IADL were assessed by interviewing participants and their caregivers [13,14]. Physical items (meal taking, bowel and bladder continence, standing ability, extent of general activities, bathing and dressing abilities), sensory items (auditory acuity and eyesight) and cognitive abilities (comprehension and self-expression) were included in the ADL. Each item was classified into five categories of self-sufficiency: completely independent, independent but slow, independent with difficulty, partially dependent and completely dependent, using a point score from 12 to 1, respectively.

The study was approved by local University Hospital Ethics Committee; the purpose and procedures of the study were explained to the subjects, and informed and written consent was obtained from the participants or caregivers.

Results

Table 1 depicts the prevalence of centenarians in Italy and in the study area; we have identified 19 centenarian, 10 female and 9 males among a population of 18,327 inhabitants. In this area the centenarian number was 4.32-fold higher than the national average (10.37 vs. 2.4/10,000). It is noteworthy that the male centenarian number was 11.51-fold higher than the national average (10.24 vs. 0.89/10,000). Female/male ratio was 1.1:1 in the study area, while the national ratio was 4.54:1.

All the centenarians live in a family home, mostly with their relatives. Individual ADL and IADL scores were in the category of moderately independent for both genders. A good anamnesis on a single individual reported a poor auditory acuity and poor eyesight, while they were free from cardiac heart disease, severe cognitive impairment, severe physical impairment, clinically evident cancer or renal insufficiency.

In Figure 1, MNA example administrated to healthy centenarian is shown [13], whereas Figure 2 reports a typical daily diet. Centenarians recruited in these area tended to be physically active, non-obese, small in stature, with a regular BMI (23.6 ± 3.1), suggestive of some degree of calorie restriction with high intake of seasonal plant food and low meat intake. Their diet shows a low glycemic index because low of refined carbohydrate (no white bread, low amount of pasta, no sweeteners, sweet beverages, can food, frozen already prepared vegetables or dishes, cookies cakes or snacks). Furthermore, they have a good intake of olive and virgin olive oil from different cultivar namely: Nocellara of Belice, Biancolilla, Giarraffa and Ogliarola that seems to have important anti-oxidant properties (unpublished data). In Figure 3, BMI and MNA are plotted together showing a perfect accordance between nutrition and body mass index.

Discussion

In Italy in 2010 the population aged one hundred years and over has 14 thousand units, hence the centenarian prevalence is 2.4/10.000. Currently there are almost five centenarian women for one man. However, in Italy there is a North to South gradient in the female/male ratio in centenarians [15]. To gain insight into the role of gender and environment, we have started a demographic study in Sicily showing that in mountain zones of Sicily there is a zone of male longevity with similar features to those found in Sardinia, in the so-called Blue Zone [11,12,16]. In both cases the municipalities concerned do not

Table 1 Distribution of Centenarian population in five villages of Sicani Mountains and in Italy

	Total Population	Males	Females	Total Centenarians	Male Centenarians	Female Centenarians
Sicani Mountains	18,328	8,793	9,535	19	9	10
	(10.37)	(10.24)	(10.48)			
Italy	60,626,442	29,413,274	31,213,168	14,473	2,612	11,861
	(2.39)	(0.89)	(3.80)			

Centenarian prevalence x 10,000 inhabitants between brackets. For data on Italian Centenarians see [15].

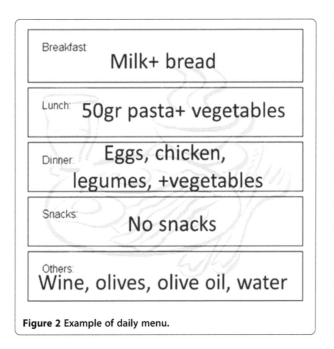

Figure 2 Example of daily menu.

include polluted areas and are small, with the lowest number of inhabitants. Therefore, longevity is more prevalent in men living in a small town, without pollution, likely because of different working conditions, different life style *i.e.* reduced smoking and alcohol abuse and Mediterranean diet. Accordingly, both these areas in Sicily and in Sardinia also share low mortality from cancers and cardiovascular diseases [11,16-18].

One of the places in the Western part of Sicily, characterized by a high presence of oldest old people, is the area of "Monti Sicani", between the provinces of Agrigento and Palermo [11,12]. "Monti Sicani" encompasses the area between the cities of Palermo and Agrigento from North to South and between the city of Caltanissetta and Trapani from West to East. The territory is characterized by a hilly area of clayey sandstone or pasture and a mountain area above 900 m, consists of pelagic limestone rocks of the Mesozoic. This area is characterized by olive tree agriculture, which tolerates a large range of soil conditions, preferring a neutral to alkaline soil type.

Looking at the national ratio of centenarian per inhabitants in these area we have found more than a four-fold increase in centenarian, and regarding male female ratio of 1.1:1 times. Since Sicilian population genetics structure is very homogeneous and in Hardy-Weinberg equilibrium [19], the explanation for these data probably resides in the environmental characteristics of the study sample.

In this area, we have found a high number of centenarians in good health, with a notable increase of male centenarians. Unequivocally, their nutritional assessment showed a high adherence to the Mediterranean nutritional profile with low glycemic index food consumed. According to the scores of ADL and IADL, centenarians of both gender demonstrated a good level of independency. They did not have any cardiac risk factors or major age related diseases (e.g. cardiac heart disease, severe cognitive impairment, severe physical impairment,

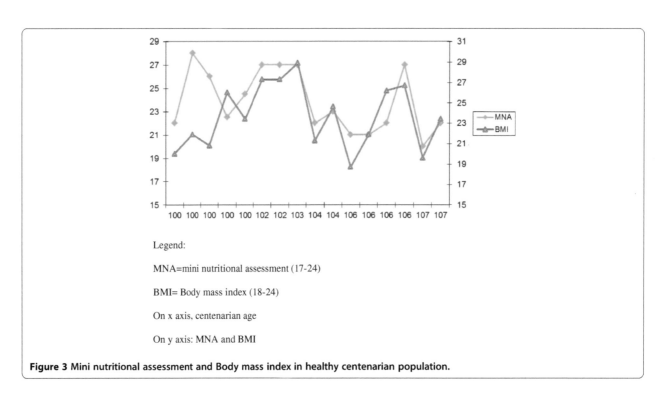

Legend:

MNA=mini nutritional assessment (17-24)

BMI= Body mass index (18-24)

On x axis, centenarian age

On y axis: MNA and BMI

Figure 3 Mini nutritional assessment and Body mass index in healthy centenarian population.

clinically evident cancer or renal insufficiency), although some had decreased auditory and visual acuity. Their life is characterized by social networking, acceptable physical activity and small amount of food divided among three meals, which contain a little amount of carbohydrate and meat and a lot of seasonal fruit and vegetables. In relation to biochemical parameters in centenarians, most biochemical parameters including cholesterol and triglycerides were within normal limits (data not shown) and better than those previously reported in a study of Sicilian elderly [20]. Furthermore, this reported modified Mediterranean-style show a low glycemic load.

The Glycemic index (GI) is defined as a kinetic parameter that reflects the potency of food to raise blood glucose level and glucose clearance. The GI of a specific diet is calculated by averaging the GI values of the food items, statistically weighted by the carbohydrate contribution. Diets based on refined carbohydrate foods that are quickly digested, absorbed, and metabolized (i.e., high glycemic index diets) have been associated with increased risk of lifestyle diseases in particular with an increased risk of type 2 diabetes, because of postprandial hyperglycemia and hyperinsulinemia related to eating high-GI carbohydrates. Low GI is known to protect against heart disease in women, and cross-sectional studies indicate low GI may reduce high-density-lipoprotein cholesterol and triacylglycerol levels in both sexes. More interesting, new observational studies have reported increased risks of coronary heart disease associated with higher intakes of carbohydrates from high glycemic index foods. Epidemiological evidence has emerged linking dietary glycemic index to visceral fat and inflammatory disease mortality [21-23]. Therefore, the Mediterranean diet is an anti-inflammatory diet [24]: it is puzzling that Italian centenarians are remarkably enriched in "good" genotypes involved in control of inflammation, confirming that a good control of inflammatory responses (genetic and/or environmental) is advantageous for longevity [25,26].

Overall, our data confirm our previous suggestion that longevity concerns subjects, living in small town, without pollution, with different working conditions, lifestyles and close adherence to a Mediterranean diet. The reason why longevity has been observed particularly in small municipalities is not surprising. It is a well established, in fact, that individuals with greater access to social support and family network have better health and lower levels of mortality, particularly when adult daughters are present. Nevertheless, our data are collected in a relative small sample of subjects; accordingly, our data needed to be confirmed by larger population-based studies.

To conclude, our work show a segment of our population that is growing faster and represent a typical example of successful ageing. Genetic and environment play a major role in healthy ageing and nutrition has a significant influence. It has been estimated that the number of centenarians will approach 3.2 million worldwide by 2050 and that means an 18-fold increase with respect to the last century [27]. Consequently, understanding the influence of dietary life-style in the process of healthy ageing is of paramount importance to development new strategies leading to healthy life extension. Finally, our results are consistent with data described in Dan Buettner's book on the importance of the diets in 5 populations with high longevity [28]. To reach successful ageing it is advisable to follow a diet with low quantity of saturated fat and high amount of fruits and vegetable, rich in phytochemicals.

Competing interests
The authors declare that they have no competing interests.

Acknowledgements
C.R. is a PhD student of PhD course in Molecular Medicine directed by C.C. and this paper is submitted in partial fulfillment of requirement for herr PhD degree. This work was supported by grants from the Ministry of Education, University and Research ex60% to C.C.- The Authors are deeply indebted with Dr. Marchese for the access to centenarians living in the villages of Bisacquino, Chiusa Sclafani and Giuliana and Dr. Pizzolanti for the information on the centenarians living in Castronovo and Prizzi.

Author details
[1]Department of Molecular and Biomolecular Sciences (STEMBIO), University of Palermo, Via Archifari 32, 90213, Palermo, Italy. [2]Department of Pathobiology and Medical and Forensic Biotechnologies (DIBIMEF), University of Palermo, Corso Tukory, 211, 90138, Palermo, Italy. [3]Immunohaemathology Unit, University Hospital, University of Palermo, Via del Vespro, 127, 90137, Palermo, Italy.

Authors' contributions
SV wrote the paper. All authors edited the paper and approved its final version.

Received: 13 April 2012 Accepted: 23 April 2012
Published: 23 April 2012

References
1. Keys A, Menotti A, Karvonen MJ, Aravanis C, Blackburn H, Buzina R, Djordjevic BS, Dontas AS, Fidanza F, Keys MH, *et al*: **The diet and 15-year death rate in the seven countries study.** *Am J Epidemiol* 1986, **124**:903–915.
2. Fidanza F, Alberti A, Lanti M, Menotti A: **Mediterranean diet score: correlation with 25-year mortality from coronary heart disease in the Seven Countries Study.** *Nutr Metab Cardiovasc Dis* 2004, **14**:254–258.
3. Sacks FM, Obarzanek E, Windhauser MM, Svetkey LP, Vollmer WM, McCullough M, Karanja N, Lin PH, Steele P, Proschan MA, *et al*: **Rationale and design of the Dietary Approaches to Stop Hypertension trial (DASH). A multicenter controlled-feeding study of dietary patterns to lower blood pressure.** *Ann Epidemiol* 1995, **5**:108–118.
4. Berry EM, Arnoni Y, Aviram M: **The Middle Eastern and biblical origins of the Mediterranean diet.** *Public Health Nutr* 2011, **14**:2288–2295.
5. Tyrovolas S, Panagiotakos DB: **The role of Mediterranean type of diet on the development of cancer and cardiovascular disease, in the elderly: a systematic review.** *Maturitas* 2010, **65**:122–130.
6. Sofi F, Cesari F, Abbate R, Gensini GF, Casini A: **Adherence to Mediterranean diet and health status: meta-analysis.** *BMJ* 2008, **337**:a1344.
7. Fung TT, Hu FB, Wu K, Chiuve SE, Fuchs CS, Giovannucci E: **The Mediterranean and Dietary Approaches to Stop Hypertension (DASH) diets and colorectal cancer.** *Am J Clin Nutr* 2010, **92**:1429–1435.
8. Bulló M, Lamuela-Raventós R, Salas-Salvadó J: **Mediterranean diet and oxidation: nuts and olive oil as important sources of fat and antioxidants.** *Curr Top Med Chem* 2011, **11**:1797–1810.

9. Aguilera CM, Mesa MD, Ramirez-Tortosa MC, Nestares MT, Ros E, Gil A: Sunflower oil does not protect against LDL oxidation as virgin olive oil does in patients with peripheral vascular disease. *Clinical Nutrition* 2010, **23**:673–681.

10. Cicerale S, Conlan XA, Sinclair AJ, Keast RS: Chemistry and health of olive oil phenolics. *Crit Rev Food Sci Nutr* 2009, **49**:218–236.

11. Bürkle A, Caselli G, Franceschi C, Mariani E, Sansoni P, Santoni A, Vecchio G, Witkowski JM, Caruso C: Pathophysiology of ageing, longevity and age related diseases. *Immun Ageing* 2007, **4**:4.

12. Vasto S, Scapagnini G, Rizzo C, Monastero R, Marchese A, Caruso C: Mediterranean diet and longevity: a survey in Sicani mountains population. *Rejuvenation research.* 2012, **15**:184–188.

13. Sancarlo D, D'Onofrio G, Franceschi M, Scarcelli C, Niro V, Addante F, Copetti M, Ferrucci L, Fontana L, Pilotto A: Prognostic Index (m-MPI) including the Mini Nutritional Assessment Short-Form (MNA-SF) for the prediction of one-year mortality in hospitalized elderly patients. *J Nutr Health Aging* 2011, **15**:169–173.

14. Lawton MP, Brody EM: Assessment of older people: self-maintaining and instrumental activities of daily living. *Gerontologist* 1969, **9**:179–186.

15. www.stat.unipd.it/ricerca/fulltext?wp=388 (in Italian)

16. Poulain M, Pes GM, Grasland C, Carru C, Ferrucci L, Baggio G, Franceschi C, Deiana L: Identification of a geographic area characterized by extreme longevity in the Sardinia island: the AKEA study. *Exp Gerontol* 2004, **39**:1423–1429.

17. Caselli G, Lipsi RM: Survival differences among the oldest old in Sardinia: who, what, where, and why? *DEMOGRAPHIC RESEARCH* 2006, **14**:267–294.

18. Marchese AG: *La quarta età tra umanesimo letterario e Biomedicina.* Palma Editrice, Palermo: Indagine sulla longevità dei Monti Sicani; 2011 (in Italian).

19. Piazza A, Olivetti E, Griffo RM, Rendine S, Amoroso A, Barbanti M, Caruso C, Conighi C, Conte R, Favoino B, *et al*: The distribution of HLA antigens in Italy. *Gene Geogr* 1989, **3**:141–164.

20. Lio D, Malaguarnera M, Maugeri D, Ferlito L, Bennati E, Scola L, Motta M, Caruso C: Laboratory parameters in centenarians of Italian ancestry. *Exp Gerontol* 2008, **43**:119–122.

21. McGeoch SC, Holtrop G, Fyfe C, Lobley GE, Pearson DW, Abraham P, Megson IL, Macrury SM, Johnstone AM: Food intake and dietary glycaemic index in free-living adults with and without type 2 diabetes mellitus. *Nutrients* 2011, **3**:683–693.

22. Brand-Miller J, Buyken AE: The glycemic index issue. *Curr Opin Lipidol* 2012, **23**:62–67.

23. Hare-Bruun H, Nielsen BM, Grau K, Oxlund AL, Heitmann BL: Should glycemic index and glycemic load be considered in dietary recommendations? *Nutr Rev* 2008, **66**:569–590.

24. Lucas L, Russell A, Keast R: Molecular mechanisms of inflammation. Anti-inflammatory benefits of virgin olive oil and the phenolic compound oleocanthal. *Curr Pharm Des* 2011, **17**:754–768.

25. Candore G, Caruso C, Colonna-Romano G: Inflammation, genetic background and longevity. *Biogerontology* 2010, **11**:565–573.

26. Vasto S, Candore G, Balistreri CR, Caruso M, Colonna-Romano G, Grimaldi MP, Listi F, Nuzzo D, Lio D, Caruso C: Inflammatory networks in ageing, age-related diseases and longevity. *Mech Ageing Dev* 2007, **128**:83–91.

27. www.un.org/spanish/esa/population/wpp2000h.pdf

28. Buettner D: *The Blue Zone: Lessons for living longer from the people who've lived the longest.* Washington, DC: National Geographic Society; 2008.

doi:10.1186/1742-4933-9-10
Cite this article as: Vasto *et al.*: Centenarians and diet: what they eat in the Western part of Sicily. *Immunity & Ageing* 2012 **9**:10.

Davinelli *et al. Immunity & Ageing* 2012, **9**:9
http://www.immunityageing.com/content/9/1/9

IMMUNITY & AGEING

REVIEW   **Open Access**

Extending healthy ageing: nutrient sensitive pathway and centenarian population

Sergio Davinelli[1], D Craig Willcox[2] and Giovanni Scapagnini[1*]

Abstract

Ageing is a challenge for any living organism and human longevity is a complex phenotype. With increasing life expectancy, maintaining long-term health, functionality and well-being during ageing has become an essential goal. To increase our understanding of how ageing works, it may be advantageous to analyze the phenotype of centenarians, perhaps one of the best examples of successful ageing. Healthy ageing involves the interaction between genes, the environment, and lifestyle factors, particularly diet. Besides evaluating specific gene-environment interactions in relation to exceptional longevity, it is important to focus attention on modifiable lifestyle factors such as diet and nutrition to achieve extension of health span. Furthermore, a better understanding of human longevity may assist in the design of strategies to extend the duration of optimal human health. In this article we briefly discuss relevant topics on ageing and longevity with particular focus on dietary patterns of centenarians and nutrient-sensing pathways that have a pivotal role in the regulation of life span. Finally, we also discuss the potential role of Nrf2 system in the pro-ageing signaling emphasizing its phytohormetic activation.

Introduction

Ageing is an irreversible process associated with numerous physiological alterations across multiple organ systems. Molecular studies in model organisms have identified several longevity genes and pathways which can extend the lifespan. Although many data are available from these animal models, in humans the situation is much more complex. Certainly human ageing is due to interactions between genetic and epigenetic factors but in addition to the genetic background, successful or unsuccessful ageing is also determined by environmental factors associated with social structure, culture and lifestyle. A fascinating, rapidly emerging concept in the biomedical sciences which may help establish a novel and innovative intellectual framework in biomedical research is the 'positive biology paradigm' [1]. Rather than making disease the central focus of researchers' efforts, positive biology seeks to understand the causes of positive phenotypes and which biological mechanisms would explain health and well-being. For instance, a better understanding of exemplars of human exceptional longevity could be a goal of positive biology and could create real benefits for those who are more vulnerable to disease and disability. Life expectancy

for humans has more than doubled in the last two centuries and in some European countries it is estimated that by 2050 the proportion of persons older than 60 will rise from 20% to almost 40% [2] and the number of centenarians will be nearly 3.2 million world-wide [3]. Therefore, even though scientists have elucidated many biological ageing processes providing new strategies that may help to slow the rate of ageing in humans, it is imperative to emphasize research on healthy ageing in order to reduce frailty and disability associated with the "normal" ageing process. Resistance against cellular stress and environmental insults can help promote a more successful ageing process that results in a longer and healthier lifespan. Despite the fact that genetic, nutritional and pharmacological interventions have been identified as potential means to slow ageing and extend lifespan in lower organisms [4], caloric restriction (CR) appears to be the only common way to increase lifespan in all species [5]. Nutrient sensors modulate lifespan extensions that occur in response to different environmental and physiological signals. Nutrient-sensing pathways are essential to the ageing process because several nutrients can activate different pathways directly or indirectly. Many of the genes that act as key regulators of lifespan also have known functions in nutrient sensing, and thus are called "nutrient-sensing longevity genes". Some examples of nutrient-sensing pathways

* Correspondence: g.scapagnini@gmail.com
[1]Department of Health Sciences, University of Molise, Campobasso, Italy
Full list of author information is available at the end of the article

involved in the longevity response are the kinase target of rapamycin (TOR) [6], AMP kinase (AMPK) [7], sirtuins [8] and insulin and insulin/insulin-like growth factor (IGF-1) signaling [9]. Among the processes and experiential factors that guide successful ageing trajectories, nutrition has been receiving much attention as a modifiable lifestyle factor leading to healthy ageing. Literature concerning the heterogeneity in dietary and nutritional status of centenarians seems to indicate that there is not any particular dietary pattern that promotes exceptional longevity. Although different nutritional compounds have been analyzed in studies of health ageing and longevity [10], it is crucial to understand how specific nutritional components and dietary patterns may affect health and longevity. To date the main dietary intervention that may retard the ageing process is CR and a rare human example could be the Okinawan population in Japan. Okinawans appear to have undergone a mild form of prolonged CR for decades that could have contributed to a lower risk of mortality [11]. A deep knowledge of the mechanisms underlying differences among centenarians from various countries would be beneficial, especially elucidating the contribution of country-specific dietary patterns. Here we will focus on specific nutritional patterns of centenarians located throughout the world considering the role of nutrient-sensing pathways in mediating the longevity response and beneficial effects. Finally, considering that CR is a mild stress that actives cytoprotective mechanisms, we discuss the potential role of Nrf2 protective cell-signaling pathway in CR induced longevity.

Nutritional patterns of centenarians: nature vs. nurture

A challenge in the area of healthy ageing is to identify dietary patterns, in addition to specific dietary components, that offer protection against age-related diseases. Dietary patterns are defined mainly for assessing eating behavior and to relate the food intake to disease or health outcomes [12]. Although severe disabilities in persons older than 60 seem to be declining, the prevalence of chronic disease, particularly of those diseases linked to diet and lifestyle, appears to be increasing [13]. Healthy centenarians are 'expert survivors' with important lessons to share, in particular regarding the most modifiable lifestyle factor: the diet. However, it is necessary to emphasize that several lines of evidence suggest that the genetic contribution to a healthy life span in populations with exceptional longevity may be greater than that seen in the general population [14,15]. A recent study on a cohort of 477 Ashkenazi Jewish centenarians with exceptional longevity reported that centenarians may possess additional longevity genes that help to buffer them against the harmful effects of an unhealthy lifestyle [16]. Nonetheless, even though studies of dietary patterns are intrinsically complex, several reports have showed that specific dietary patterns are potentially associated with longevity [17,18]. The diets of 5 populations with extraordinarily high longevity have been recently described and labeled "Blue Zones". [19]. Populations of Okinawa, Japan; Sardinia, Italy; Loma Linda, California; the Nicoya Peninsula, Costa Rica and Ikaria, Greece seem to have a high prevalence of centenarians and a preferential attitude toward a plant based-diet. Interestingly, in 2003 Shimizu et al. investigated the dietary practices of 104 centenarians who lived in the Tokyo metropolitan area and reported that a dietary pattern based on dairy products was associated with increased survival [20]. Traditional Okinawan diets provide about 90% of calories from carbohydrates but in vegetable form [21] therefore are low in calories but nutritionally dense, particularly with regard to vitamins, minerals, and phytonutrients [18]. Okinawans also have a very high intake of phytochemicals in the diet. All plants contain these natural compounds and the elders seem to have significantly lower levels of lipid peroxidation and suffer less free-radical-induced damage. For many individuals, the cognitive changes that occur with ageing are affected by micronutrient intake. Recently, a pilot study compared circulating levels of micronutrients among cognitively healthy volunteers aged 85 years and older in Okinawa and Oregon. The Okinawan elders used fewer vitamin supplements but had similar levels of vitamin B12 and α-tocopherol, compared with Oregonian elders. Thus the components leading to healthy cognitive ageing might include a variety of patterns that include a healthy diet, high physical activity, and social engagement [22]. Additionally, cognitive function, daily activity, and residential status, have been reported to affect nutritional intake of centenarians [23,24]. The traditional Mediterranean diet provides about 40% of calories from fat, mostly monounsaturated and polyunsaturated fat [21]. The benefits of a Mediterranean diet are well known but a specific region in the Mediterranean island of Sardinia ('Blue Zone') is characterized by exceptional male longevity. Noteworthy, an analysis in the 377 Sardinian municipalities provided evidence that dietary variables are not significantly correlated with male extreme longevity. In particular, it was revealed that a lower caloric intake is not related with a superior level of longevity but rather factors affecting energy expenditure are important in explaining extreme longevity [25]. These findings seem to indicate that in this area CR had an impact, although minor on exceptional longevity. In Loma Linda, California, there is a community of Seventh Day Adventists which, according to several studies, live longer than the rest of the population. Interestingly, their vegetarian diet is thought to be the most likely cause of their extraordinary longevity. Specific dietary factors that may be involved in their outstanding health include a high intake of fruit, vegetables, and nuts [26-28]. The Nicoya Peninsula region in Costa Rica has been reported to be an exceptional longevity area where healthy centenarians live surrounded by a solid support network of friends and family [29]. Although the possible role played by the dietary regimen in Nicoya region in relation to extreme longevity has not yet been investigated, the diet includes garden vegetables, an abundance of fruit (orange, mango, papaya), squash, beans, rice and corn. The water is also particularly high in minerals such as magnesium and calcium [19]. This area has also reported one of the lowest middle-age mortality rates in the world. A 60-year -old has more than a fourfold better chance of making it to the age of 90 years than a 60-year-old in North America [30]. Finally, it was reported that people in Ikaria Island, Greece, have also one of the highest life expectancies in the world. Ikarians are three times more likely to reach the age of 90 years than in the U.S. [19]. In this community, scientific evidence shows protective health benefits from long-term adherence to the Mediterranean food culture revealing that this diet has a cardioprotective effect and is able to reduce the prevalence of hyperuricaemia in elderly individuals [31]. The above mentioned regions have been labeled as 'Blue Zones' and while scientists try to validate the veracity and variety of associated causes of this exceptional longevity, it is advisable to follow a diet rich in fruits, vegetables, legumes and whole grains but reduced in saturated fat.

'Nutrient sensors' that modulate ageing

Biochemical pathways capable of 'sensing' the availability of nutrients maintain energy homeostasis both at the cell and at the whole organism levels [32]. Multiple nutrient signaling pathways have been connected to lifespan regulation. Although an extremely low-calorie diet is the most effective intervention known to extend lifespan in many species, from yeast to primates [33], we will focus on nutrient-sensing pathways that have been shown to influence ageing in humans. The full understanding of dietary intake and composition to obtain health benefits and pro-longevity effects is a realistic goal for biogerontology and drugs that promote human longevity targeting nutrient-sensing pathways require further study. Moreover, it has been observed that natural genetic variants in nutrient-sensing pathways are associated with increased human life span [34]. In recent years, some of these pathways identified in worms or mice have been shown to have human homologs, in particular the IGF-1 pathway. IGF-1 activity is essential in all animals, and surprisingly, altered IGF-1 signaling pathways have been shown to confer an increase in susceptibility to longevity, including human longevity. Recent studies have demonstrated a significant association between mutations in genes involved in the IGF-1 pathway and extension of human

life span. Indeed mutations known to impair IGF-1 receptor function have been shown to be overrepresented in a cohort of Ashkenazi Jewish centenarians suggesting that centenarians may harbor rare genetic variations in genes encoding components of the IGF-1 pathway [35]. Polymorphic variants of genes which are involved in IGF-1 signaling have also been linked to longevity in a Japanese cohort of 122 semi-supercentenarians (105 years old and older) [36]. The FOXO transcription factor FOXO3, part of the IGF-1 pathway, is essential for CR effects and it has been demonstrated that polymorphisms in FOXO3 are associated with human longevity in several cohorts. Specifically, building upon previous findings in animal models, Willcox et al. (2008) identified a strong association between FOXO3A and human longevity in Japanese-Americans from Hawaii [37]. Subsequently, Pawlikowska et al. reported common variants in both FOXO3A and AKT1, to be associated with longer lifespan in three independent Caucasian cohorts [38]. Anselmi et al. also validated the association between FOXO3A polymorphisms and extreme longevity in males from the southern Italian Centenarian Study [39]. Moreover, the key role of FOXO3 in human longevity has been further confirmed in German and Chinese centenarians [40,41]. However, despite the recognition of FOXO3 as a "master gene" in ageing the actual functional variant remains unidentified. Recently, Donlon et al. sequenced the coding region in their long-lived Japanese-American population demonstrating that of 38 published variants the vast majority remained unconfirmed, indicating that coding variants may not be key players [42]. Enhanced resources for fine-mapping this region are necessary. The TOR (target of rapamycin) pathway is an evolutionarily conserved nutrient-sensing pathway which has been implicated in the regulation of life span and in the response to stress, nutrients and growth factors [43]. As Blagosklonny argued, ageing may not necessarily be driven by damage, but, in contrast, lead- to damage and this process is driven in part by mTOR (mammalian target of rapamycin) [44]. However, to date human data have been scarce and the details of how mTOR exerts lifespan control and anti-ageing effects are still not fully understood. Notably, it was recently demonstrated that rapamycin reverses the phenotype of cells obtained from patients with Hutchinson-Gilford progeria syndrome, a lethal genetic disorder that mimics rapid ageing [45]. Taking into account that mTOR signaling is a major nutrient-sensing pathway with effects on ageing, the inhibition of this signaling pathway may be similar to what is seen in dietary restriction and, as discussed earlier, is known to extend the lifespan of diverse organisms [5]. Maintenance of mitochondrial activity is also an emerging topic in the field of ageing research and many reports indicate that an effective mitochondrial fitness is essential for healthy ageing [46]. AMPK is a nutrient and energy sensor that might be involved in the regulation of life span and in the mediation of the beneficial effects of CR. Although this hypothesis is largely unexplored, especially in mammals, it seem likely that the activation of AMPK may have an impact on the activity of FOXO, sirtuin, and the mTOR pathways, which have been linked to CR and to the promotion of a healthy longevity [47].

Hormetic phytochemicals and Nrf2- signaling pathway in healthy ageing

There is a huge volume of data on the beneficial effects of plant-derived extracts to retard ageing and age-associated diseases. Many phytochemicals are synthesized to increase the fitness of the plant by allowing it to interact with its environment, including herbivorous pathogens, and insects. Some well-known phytochemicals may induce in humans beneficial stress responses and with low-dose exposures, they can trigger a cellular stress response and subsequently induce adaptive stress resistance, also called hormesis [48]. Stress resistance involves several molecular adaptations and induces many of the nutrient-sensing longevity pathways discussed above. The role of hormesis in ageing has already been examined by Rattan [49] but also its relevance to explain the anti-ageing and life-extending actions of CR in long-lived species [50]. We propose that the anti-ageing responses induced by phytochemicals are caused by phytohormetic stress resistance involving the activation of Nrf2 signaling. There is a substantial amount of research supporting oxidative stress as one of the main causes of ageing. In contrast there are few longevity models that have been created to evaluate enhanced anti-oxidative mechanisms. Nrf2 is a central regulator of the adaptive response to oxidative stress but very few studies have investigated the role of Nrf2 in the modulation of ageing and longevity. The Nrf2-signaling pathway has been extensively reviewed elsewhere [51,52] as well as its hormetic function from an evolutionary perspective [53]. It is important to delineate the link between oxidative stress, cellular resistance and the rate of ageing. Since Nrf2 regulates the main cytoprotective responses, this pathway may substantially contribute to the determination of healthspan and extension of longevity (Figure 1). Moreover, many phytochemicals (e.g. polyphenols, flavonoids, terpenoids, etc.) are major ingredients present in fruits, vegetables, and spices and have been shown to have protective effects- against age-related degeneration [54]. Interestingly, many of these phytochemicals are activators of Nrf2 signaling and through this pathway they can inhibit ROS production and counteract oxidative damage [51]. Furthermore, hormetic phytochemicals have recently received considerable attention for their pro-

longevity effects and for their ability to act as sirtuin activators [55]. Taking into account that the dietary habits of many centenarians seem to be extremely rich in phytochemicals, without neglecting the effect of genetic background, we hypothesize that long-lived people may have a constitutively upregulated Nrf2 pathway to respond better to cell stressors and thereby minimize cell damage. Moreover, considering that so-called "Blue Zones" that seem to harbor a high number of centenarians appear also to be areas with high nutrient density and low caloric density diets which may have led to a prolonged form of mild CR, the beneficial effects of their CR exposure may be also partially explained by a decrease in oxidative stress sensitivity, suggesting Nrf2 as a plausible effector of longevity signaling. Finally, it is important to consider that emerging evidence shows the close link between nutrient intake and inflammatory biomarkers [56]. It is well established that adipose tissue releases many inflammatory mediators and experimental studies have revealed the beneficial effects of CR in the attenuation of system-wide inflammatory processes [57]. Moreover, oxidative stress has been recognized to play a major role in determining and maintaining the low grade state of inflammation observed in ageing and age associated diseases [58]. Consistent with these statements, a number of studies have shown that the antioxidant-mediated Nrf2 activation is strongly associated with the protection from pro-inflammatory insults [59]. The activation of Nrf2 pathway might inhibit the production or expression of pro-inflammatory mediators including cytokines, chemokines, cell adhesion molecules, matrix metalloproteinases, cyclooxygenase-2 and inducible nitric oxide synthase [60]. Therefore, efficient inducers of Nrf2 activation, some of which are present in the diet of centenarians could be considered as effective means for the prevention of inflammation-mediated diseases.

Conclusions

This review summarized some of the healthy ageing secrets of long-lived individuals. Presently it is of great interest to study characteristics of people living over far longer than the expected life span and to understand which factors are important in shaping longevity. The realization of healthy longevity is possible but to achieve a longer and a healthier life, increased attention must be placed on lifestyle choices, particularly the diet. There is a huge volume of scientific literature on diet and health but less attention has been paid to dietary patterns. Although it seems unlikely that there is a particular dietary pattern that promotes exceptional longevity, understanding the heterogeneity in dietary patterns and nutritional status of centenarians may provide a wealth of information relevant to human ageing. From a scientific perspective, a

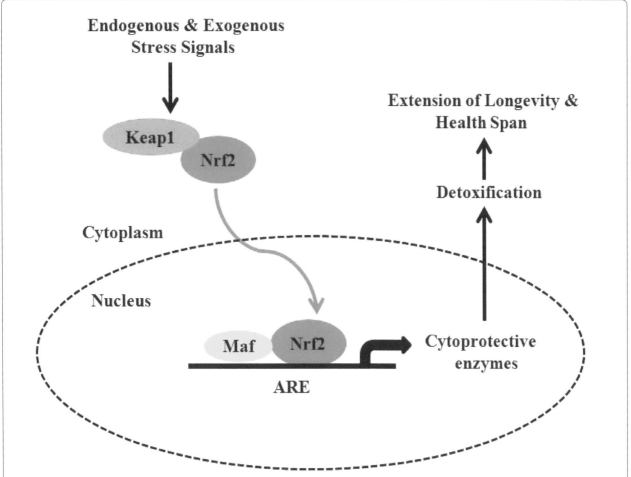

Figure 1 General scheme for the induction of Nrf2-signaling pathway. The antioxidant response element (ARE) in the promoter region of select genes allows the coordinated up-regulation of antioxidant and detoxifying enzymes in response to dietary phytochemicals. This up-regulation is mediated through nuclear factor (erythroid-derived 2)-like 2 (Nrf2) that may be activated by endogenous and exogenous molecules or stressful conditions. These agents disrupt the association between Nrf2 and Keap1 with subsequent nuclear translocation of Nrf2. In the cell nucleus Nrf2 interacts with small MAF protein, forming a heterodimer that binds to the ARE sequence in the promoter region and up-regulates transcription of many genes encoding detoxifying enzymes. We speculate that this signaling pathway is constitutively upregulated in long-lived individuals providing extension of longevity and health span.

particular diet able to delay ageing may help to identify new molecules to extend and ameliorate age associated disease, opening new opportunities for drug discovery and companies working in nutrition and pharmacology. Furthermore, our knowledge of nutrient-sensing pathways has greatly increased in recent years and the modulation of these pathways by diet or pharmaceuticals can have a profound impact on health and thus represent a therapeutic opportunity for the extension of the human lifespan and quality of life improvement. In addition to the common pathways that regulate biological ageing there are also promising and attractive new targets for therapeutic interventions that can positively affect healthy ageing. The development of strategies that will lead to the extension of healthy life and that would result in slowing the rate of ageing and lowering risk for age associated disease may be part of the new paradigm for the biomedical sciences that can be termed 'positive biology'.

Acknowledgements
Dr. D. Craig Willcox gratefully acknowledges the support of the Kuakini Hawaii Healthspan Study (1R01AG038707-01A1) and the Kuakini Hawaii Lifespan II Study (5R01AG027060-06).

Author details
[1]Department of Health Sciences, University of Molise, Campobasso, Italy.
[2]Department of Human Welfare, Okinawa International University, Ginowan, Japan.

Authors' contributions
SD, DCW, GS wrote the draft; all authors edited the paper and approved its final version.

Competing interests
The authors declare that they have no competing interests.

Received: 3 April 2012 Accepted: 23 April 2012 Published: 23 April 2012

References

1. Farrelly C: 'Positive biology' as a new paradigm for the medical sciences. Focusing on people who live long, happy, healthy lives might hold the key to improving human well-being. *EMBO Rep* 2012, 13:186-188.
2. Weiss G: **Europe wakes up to aging.** *Sci Aging Knowledge Environ* 2002, **48**:ns10.
3. World population projections: the 2000 revision. Population Division Department of Economic and Social Affairs, United Nations. [http://www.un.org/spanish/esa/population/wpp2000h.pdf].
4. Vijg J, Campisi J: **Puzzles, promises and a cure for ageing.** *Nature* 2008, **454**:1065-1071.
5. Bishop NA, Guarente L: **Genetic links between diet and lifespan: shared mechanisms from yeast to humans.** *Nat Rev Genet* 2007, **8**:835-844.
6. Hansen M, Taubert S, Crawford D, Libina N, Lee SJ, Kenyon C: **Lifespan extension by conditions that inhibit translation in Caenorhabditis elegans.** *Aging Cell* 2007, **6**:95-110.
7. Greer EL, Dowlatshahi D, Banko MR, Villen J, Hoang K, Blanchard D, Gygi SP, Brunet A: **An AMPK-FOXO pathway mediates longevity induced by a novel method of dietary restriction in C. elegans.** *Curr Biol* 2007, **17**:1646-1656.
8. Li Y, Xu W, McBurney MW, Longo VD: **SirT1 inhibition reduces IGF-I/IRS-2/Ras/ERK1/2 signaling and protects neurons.** *Cell Metab* 2008, **8**:38-48.
9. Honjoh S, Yamamoto T, Uno M, Nishida E: **Signalling through RHEB-1 mediates intermittent fasting-induced longevity in C. elegans.** *Nature* 2009, **457**:726-730.
10. Lebel M, Picard F, Ferland G, Gaudreau P: **Drugs, nutrients, and phytoactive principles improving the health span of rodent models of human age-related diseases.** *J Gerontol A Biol Sci Med Sci* 2012, **67**:140-151.
11. Willcox DC, Willcox BJ, Todoriki H, Curb JD, Suzuki M: **Caloric restriction and human longevity: what can we learn from the Okinawans?** *Biogerontology* 2006, **7**:173-177.
12. Román-Viñas B, Ribas Barba L, Ngo J, Martínez-González MA, Wijnhoven TM, Serra-Majem L: **Validity of dietary patterns to assess nutrient intake adequacy.** *Br J Nutr* 2009, **101**:(Suppl 2):S12-S20.
13. National Center for Health Statistics: *Health, United States, 2004 with chartbook of trends in the health of Americans* Hyattsville, MD: National Center for Health Statistics; 2004.
14. Perls TT, Wilmoth J, Levenson R, Drinkwater M, Cohen M, Bogan H, Joyce E, Brewster S, Kunkel L, Puca A: **Life-long sustained mortality advantage of siblings of centenarians.** *Proc Natl Acad Sci USA* 2002, **99**:8442-8447.
15. Sebastiani P, Solovieff N, Puca A, Hartley SW, Melista E, Andersen S, Dworkis DA, Wilk JB, Myers RH, Steinberg MH, Montano M, Baldwin CT, Perls TT: **Genetic signatures of exceptional longevity in humans.** *Science* 2010, **333**:404.
16. Rajpathak SN, Liu Y, Ben-David O, Reddy S, Atzmon G, Crandall J, Barzilai N: **Lifestyle factors of people with exceptional longevity.** *J Am Geriatr Soc* 2011, **59**:1509-1512.
17. Willcox DC, Willcox BJ, Todoriki H, Suzuki M: **The Okinawan diet: health implications of a low-calorie, nutrient-dense, antioxidant-rich dietary pattern low in glycemic load.** *J Am Coll Nutr* 2009, **28**:500S-516S.
18. Willcox BJ, Willcox DC, Todoriki H, Fujiyoshi A, Yano K, He Q, Curb JD, Suzuki M: **Caloric restriction, the traditional Okinawan diet, and healthy aging. The diet of the world's longest-lived people and its potential impact on morbidity and life span.** *Ann N Y Acad Sci* 2007, **1114**:434-455.
19. Buettner D: *The blue zone: lessons for living longer from the people who've lived the longest* Washington, DC: National Geographic Society; 2008.
20. Shimizu K, Takeda S, Noji H, Hirose N, Ebihara Y, Arai Y, Hamamatsu M, Nakazawa S, Gondo Y, Konishi K: **Dietary patterns and further survival in Japanese centenarians.** *J Nutr Sci Vitaminol (Tokyo)* 2003, **49**:133-138.
21. Kromhout D: **Food consumption patterns in the Seven Countries Study. Seven Countries Study Research Group.** *Ann Med* 1989, **21**:237-238.
22. Dodge HH, Katsumata Y, Todoriki H, Yasura S, Willcox DC, Bowman GL, Willcox B, Leonard S, Clemons A, Oken BS, Kaye JA, Traber MG: **Comparisons of plasma/serum micronutrients between Okinawan and Oregonian elders: a pilot study.** *J Gerontol A Biol Sci Med Sci* 2010, **65**:1060-1067.
23. Arai Y, Hirose N, Nakazawa S, Yamamura K, Shimizu K, Takayama M, Ebihara Y, Osono Y, Homma S: **Lipoprotein metabolism in Japanese**

24. centenarians: effects of apolipoprotein E polymorphism and nutritional status. *J Am Geriatr Soc* 2001, **49**:1434-1441.
24. Johnson MA, Davey A, Hausman DB, Park S, Poon LW, the Georgia Centenarian Study: **Dietary differences between centenarians residing in communities and in skilled nursing facilities: the Georgia Centenarian Study.** *Age* 2000, **28**:333-341.
25. Pes GM, Tolu F, Poulain M, Errigo A, Masala S, Pietrobelli A, Battistini NC, Maioli M: **Lifestyle and nutrition related to male longevity in Sardinia: an ecological study.** *Nutr Metab Cardiovasc Dis* 2011.
26. Fraser GE: **Diet as primordial prevention in Seventh-Day Adventists.** *Prev Med* 1999, **29**:S18-S23.
27. Willett W: **Lessons from dietary studies in Adventists and questions for the future.** *Am J Clin Nutr* 2003, **78**:539S-543S.
28. Rizzo NS, Sabaté J, Jaceldo-Siegl K, Fraser GE: **Vegetarian dietary patterns are associated with a lower risk of metabolic syndrome: the adventist health study 2.** *Diabetes Care* 2011, **34**:1225-1227.
29. Koch T, Power C, Kralik D: **Researching with centenarians.** *Int J Older People Nurs* 2007, **2**:52-61.
30. Christensen K, Vaupel JW: **Determinants of longevity: genetic, environmental and medical factors.** *J Intern Med* 1996, **240**:333-341.
31. Chrysohoou C, Skoumas J, Pitsavos C, Masoura C, Siasos G, Galiatsatos N, Psaltopoulou T, Mylonakis C, Margazas A, Kyvelou S, Mamatas S, Panagiotakos D, Stefanadis C: **Long-term adherence to the Mediterranean diet reduces the prevalence of hyperuricaemia in elderly individuals, without known cardiovascular disease: the Ikaria study.** *Maturitas* 2011, **70**:58-64.
32. Rossetti L: **Perspective: hexosamines and nutrient sensing.** *Endocrinology* 2000, **141**:1922-1925.
33. Colman RJ, Anderson RM, Johnson SC, Kastman EK, Kosmatka KJ, Beasley TM, Allison DB, Cruzen C, Simmons HA, Kemnitz JW, Weindruch R: **Caloric restriction delays disease onset and mortality in rhesus monkeys.** *Science* 2009, **325**:201-204.
34. Bonafè M, Barbieri M, Marchegiani F, Olivieri F, Ragno E, Giampieri C, Mugianesi E, Centurelli M, Franceschi C, Paolisso G: **Polymorphic variants of insulin-like growth factor I (IGF-I) receptor and phosphoinositide 3-kinase genes affect IGF-I plasma levels and human longevity: cues for an evolutionarily conserved mechanism of life span control.** *J Clin Endocrinol Metab* 2003, **88**:3299-3304.
35. Suh Y, Atzmon G, Cho MO, Hwang D, Liu B, Leahy DJ, Barzilai N, Cohen P: **Functionally significant insulin-like growth factor I receptor mutations in centenarians.** *Proc Natl Acad Sci USA* 2008, **105**:3438-3442.
36. Kojima T, Kamei H, Aizu T, Arai Y, Takayama M, Nakazawa S, Ebihara Y, Inagaki H, Masui Y, Gondo Y, Sakaki Y, Hirose N: **Association analysis between longevity in the Japanese population and polymorphic variants of genes involved in insulin and insulin-like growth factor 1 signaling pathways.** *Exp Gerontol* 2004, **39**:1595-1598.
37. Willcox BJ, Donlon TA, He Q, Chen R, Grove JS, Yano K, Masaki KH, Willcox DC, Rodriguez B, Curb JD: **FOXO3A genotype is strongly associated with human longevity.** *Proc Natl Acad Sci USA* 2008, **105**:13987-13992.
38. Pawlikowska L, Hu D, Huntsman S, Sung A, Chu C, Chen J, Joyner AH, Schork NJ, Hsueh WC, Reiner AP, Psaty BM, Atzmon G, Barzilai N, Cummings SR, Browner WS, Kwok PY, Ziv E, Study of Osteoporotic Fractures: **Association of common genetic variation in the insulin/IGF1 signaling pathway with human longevity.** *Aging Cell* 2009, **8**:460-472.
39. Anselmi CV, Malovini A, Roncarati R, Novelli V, Villa F, Condorelli G, Bellazzi R, Puca AA: **Association of the FOXO3A locus with extreme longevity in a southern Italian centenarian study.** *Rejuvenation Res* 2009, **12**:95-104.
40. Flachsbart F, Caliebe A, Kleindorp R, Blanché H, von Eller-Eberstein H, Nikolaus S, Schreiber S, Nebel A: **Association of FOXO3A variation with human longevity confirmed in German centenarians.** *Proc Natl Acad Sci USA* 2009, **106**:2700-2705.
41. Li Y, Wang WJ, Cao H, Lu J, Wu C, Hu FY, Guo J, Zhao L, Yang F, Zhang YX, Li W, Zheng GY, Cui H, Chen X, Zhu Z, He H, Dong B, Mo X, Zeng Y, Tian XL: **Genetic association of FOXO1A and FOXO3A with longevity trait in Han Chinese populations.** *Hum Mol Genet* 2009, **18**:4897-4904.
42. Donlon TA, Curb JD, He Q, Grove J, Masaki KH, Rodriguez B, Elliot A, Willcox DC, Willcox BJ: **FOXO3 gene variants and human aging: coding variants may not be key players.** *J Gerontol A Biol Sci Med Sci* 2012.

43. Sengupta S, Peterson TR, Sabatini DM: **Regulation of the mTOR complex 1 pathway by nutrients, growth factors, and stress.** *Mol Cell* 2010, **40**:310-322.
44. Blagosklonny MV: **Why human lifespan is rapidly increasing: solving "longevity riddle" with "revealed-slow-aging" hypothesis.** *Aging (Albany NY)* 2010, **2**:177-182.
45. Cao K, Graziotto JJ, Blair CD, Mazzulli JR, Erdos MR, Krainc D, Collins FS: **Rapamycin reverses cellular phenotypes and enhances mutant protein clearance in Hutchinson-Gilford progeria syndrome cells.** *Sci Transl Med* 2011, **3**, 89ra58.
46. Lopez-Lluch G, Irusta PM, Navas P, de Cabo R: **Mitochondrial biogenesis and healthy aging.** *Exp Gerontol* 2008, **43**:813-819.
47. Salminen A, Kaarniranta K: **AMP-activated protein kinase (AMPK) controls the aging process via an integrated signaling network.** *Ageing Res Rev* 2012, **11**:230-241.
48. Mattson MP: **Dietary factors, hormesis and health.** *Ageing Res Rev* 2008, **7**:43-48.
49. Rattan SI: **Hormesis in aging.** *Ageing Res Rev* 2008, **7**:63-78.
50. Masoro EJ: **The role of hormesis in life extension by dietary restriction.** *Interdiscip Top Gerontol* 2007, **35**:1-17.
51. Scapagnini G, Vasto S, Abraham NG, Caruso C, Zella D, Fabio G: **Modulation of Nrf2/ARE pathway by food polyphenols: a nutritional neuroprotective strategy for cognitive and neurodegenerative disorders.** *Mol Neurobiol* 2011, **44**:192-201.
52. Sykiotis GP, Bohmann D: **Stress-activated cap'n'collar transcription factors in aging and human disease.** *Sci Signal* 2010, **3**:re3.
53. Maher J, Yamamoto M: **The rise of antioxidant signaling-the evolution and hormetic actions of Nrf2.** *Toxicol Appl Pharmacol* 2010, **244**:4-15.
54. Kim J, Lee HJ, Lee KW: **Naturally occurring phytochemicals for the prevention of Alzheimer's disease.** *J Neurochem* 2010, **112**:1415-1430.
55. Calabrese V, Cornelius C, Dinkova-Kostova AT, Iavicoli I, Di Paola R, Koverech A, Cuzzocrea S, Rizzarelli E, Calabrese EJ: **Cellular stress responses, hormetic phytochemicals and vitagenes in aging and longevity.** *Biochim Biophys Acta* 2012, **1822**:753-783.
56. Galland L: **Diet and inflammation.** *Nutr Clin Pract* 2010, **25**:634-640.
57. Morgan TE, Wong AM, Finch CE: **Anti-inflammatory mechanisms of dietary restriction in slowing aging processes.** *Interdiscip Top Gerontol* 2007, **35**:83-97.
58. De la Fuente M, Miquel J: **An update of the oxidation-inflammation theory of aging: the involvement of the immune system in oxi-inflamm-aging.** *Curr Pharm Des* 2009, **15**:3003-3026.
59. Khor TO, Yu S, Kong AN: **Dietary cancer chemopreventive agents - targeting inflammation and Nrf2 signaling pathway.** *Planta Med* 2008, **74**:1540-1547.
60. Reuter S, Gupta SC, Chaturvedi MM, Aggarwal BB: **Oxidative stress, inflammation, and cancer: How are they linked?** *Free Radic Biol Med* 2010, **49**:1603-1616.

doi:10.1186/1742-4933-9-9
Cite this article as: Davinelli *et al.*: **Extending healthy ageing: nutrient sensitive pathway and centenarian population.** *Immunity & Ageing* 2012 **9**:9.

Conclusion

In a heterogeneous population, such as the human population, the ability to maintain an adequate response to stressors within a range compatible with a state of good health should have a Gaussian distribution; hence, centenarians should be the extreme tail of this curve, represented by the individuals likely to maintain an adequate response to stressors and to repair the damage. Of course, they are the individuals better adapted to environmental conditions that have occurred. In the generation under study, these factors are as likely to be represented by inflammatory age-related diseases as cardiovascular ones.

The careful phenotyping of numerous animal models and aged human beings, the collection of genetic material, and the current "explosion" of molecular genetic techniques and data will soon add important missing pieces to the puzzle of longevity. However, complex gene–gene and gene–environment interactions will make it difficult to understand how genes affect ageing and longevity. Nevertheless, with demographic selection, centenarians have already been useful in deciphering some polymorphisms and genetic loci associated or not with exceptional longevity. Centenarians are thus a unique model to address those questions and doubts about the biology of human ageing that would otherwise be difficult to interpret.

Success in increasing longevity in laboratory organisms has shown that ageing is not an immutable process. Hence, the time has come to get more serious about the effort to slow human ageing or to age successfully. On the other hand, if ageing is combined with extended years of healthy life, it could also produce unprecedented social, economic, and health dividends. Understanding of the ageing process should be a crucial component of such an approach and must have a prominent role in new strategies for extending the health of a population that is highly susceptible to the diseases of ageing. Accordingly, as extensively discussed in this book, investigating longevity, particularly disentangling the role of genetics and lifestyle, is likely to provide important clues about how to develop drugs that can slow or delay ageing. Thus it will be possible to understand, through a "positive biology" approach, how to prevent and/or reduce elderly frailty and disability.

Calogero Caruso and Sonya Vasto
University of Palermo, Italy

Ingram Content Group UK Ltd.
Milton Keynes UK
UKHW052226240323
419151UK00002B/53